# Synthesis Lectures on Engineering, Science, and Technology

The focus of this series is general topics, and applications about, and for, engineers and scientists on a wide array of applications, methods and advances. Most titles cover subjects such as professional development, education, and study skills, as well as basic introductory undergraduate material and other topics appropriate for a broader and less technical audience.

Stanislaw Raczynski

# How Circuits Work

## Amplifiers, Filters, Audio and Control Electronics

Springer

Stanislaw Raczynski
Facultad de Ingenieria
Panamerican University
Mexico City, Mexico

ISSN 2690-0300 ISSN 2690-0327 (electronic)
Synthesis Lectures on Engineering, Science, and Technology
ISBN 978-3-031-34936-2 ISBN 978-3-031-34934-8 (eBook)
https://doi.org/10.1007/978-3-031-34934-8

© The Editor(s) (if applicable) and The Author(s), under exclusive license to Springer Nature Switzerland AG 2023

This work is subject to copyright. All rights are solely and exclusively licensed by the Publisher, whether the whole or part of the material is concerned, specifically the rights of translation, reprinting, reuse of illustrations, recitation, broadcasting, reproduction on microfilms or in any other physical way, and transmission or information storage and retrieval, electronic adaptation, computer software, or by similar or dissimilar methodology now known or hereafter developed.
The use of general descriptive names, registered names, trademarks, service marks, etc. in this publication does not imply, even in the absence of a specific statement, that such names are exempt from the relevant protective laws and regulations and therefore free for general use.
The publisher, the authors, and the editors are safe to assume that the advice and information in this book are believed to be true and accurate at the date of publication. Neither the publisher nor the authors or the editors give a warranty, expressed or implied, with respect to the material contained herein or for any errors or omissions that may have been made. The publisher remains neutral with regard to jurisdictional claims in published maps and institutional affiliations.

This Springer imprint is published by the registered company Springer Nature Switzerland AG
The registered company address is: Gewerbestrasse 11, 6330 Cham, Switzerland

# Preface

While writing research or textbooks, the general question is: Do we need books at all, taking into account the huge amount of information, available on the Web? Now, you can find everything using fast searching software. However, this way to get information may be somewhat difficult and confusing for a beginner student. For example, the term "Electric circuit" produces 345,000,000 results in Google. if we search for "transistor", we get 139,000,000 related pages. Most of these sources provide useful information, though many of them are superficial, or, on the other hand, contain excessive theoretical or technical detail.

The purpose of this book is to provide comprehensive and compact information on the concepts and devices in the field of basic electronics. Most used and important circuits are discussed, with an explanation of their functioning, from the practical point of view. However, to understand how circuits work, certain general theoretical background and knowledge are needed. So, the book contains such concepts as complex numbers, Fourier and Laplace transform and Z-transform. Though these are high mathematics concepts, they are used here in the most simplified form. For example, the Laplace transform is used here as a tool that simplifies and not complicates things. In fact, the only thing that the beginner really should understand is that the "s" variable of the Laplace transform can be treated as the differentiating operator. This allows us to define the transfer function of linear devices. As for the discrete-time, digital signal processing, the Z-transform is used. However, the only relevant property of the transform used here is the relation between the one-period delay and $z^{-1}$ operator.

For many of the discussed circuits, the computer simulations of the transient processes and the frequency response are shown.

The book does not include (with some little exceptions) references to commercial devices like transistors and integrated circuits. The information contained here may help the reader to find commercial ICs and other devices, as well as detailed descriptions in commonly accessible data sheets.

The book starts with the elemental concepts of electric charge, current and voltage (Chap. 1). The methods of circuit analysis are mentioned. The properties of resistors, capacitors, inductors, transformers and others, are explained. The basic electronic elements are explained, including diodes, transistors, thyristors and other mostly used devices.

In Chap. 2 we discuss the transistor-based voltage amplifiers, including power amplifiers of class A, B, C and D. Practical hints and remarks are provided.

Chapter 3 "Filters" starts with the concepts of circuit dynamics. Then, a number of signal processing circuits called *filters* are explained, including passive RC and RLC circuits, and the first- and second-order active filters based on operational amplifiers. Then, the higher order filters are discussed, using as examples the Chebyshev and Butterworth filters.

Oscillating circuits and signal generators are discussed in Chap. 4. The phase-shift, Colpittz, Hartley, Wien bridge and twin T oscillators are explained. The mono-stable, relaxation, and multivibrator circuits are discussed.

Remarks on voltage and supply power sources can be found in Chap. 5. The difference between ideal and real supply sources is mentioned, and the practical implementations are shown. Voltage regulators are considered, and related circuits, like zero-crossover, are included.

Chapter 6 is an introduction to electric control circuits. It contains some remarks on feedback control systems and the electrical realization of the two-pónt and continuous PID controller. Various versions of electric PID controllers are discussed.

The elemental concepts of the digital signal processing can be found in Chap. 7. The Z-transform is explained in its most simplified form. The Z-transfer function is used and the relation to the continuous version of the devices is discussed. Such problems as aliasing and frequency response are mentioned.

Finally, in Chap. 8 "Miscellaneous", several useful applications of integrated circuits and logical gates are included. Basic types of logic gates are discussed, and the circuits like adder, pulse generator, counter and others are discussed.

Mexico City, Mexico Stanislaw Raczynski

# Contents

# Electric Circuits and Devices 1

## 1.1 Electric Current and Voltage

### 1.1.1 Electric Charge, Current and the Law of Ohm

From the very beginning of the fascinating history of knowledge about the electricity, the researchers have been looking for the ways to measure the amount of electric charge. Looking for the definition of electric charge on the Web, most probably you will find that the charge of one Coulomb is the amount of charge accumulated by 1 A of electric current that flows during one second. If you want to see what one Amper is, the definition provided is one Coulomb per second. So, let us recall the proper definition of the unit of electric charge. It is derived from the law of Charles Coulomb (1736–1806). For more information about electric circuits consult, for example, Alexander [2].

Consider two particles located at one meter from each other, equally positively charged. The charge of each particle is equal to one *Coulomb* if the repelling force between the particles is equal to $9 \times 10^{-9}$ N. One Coulomb of electric charge is equivalent to $6{,}241 \times 10^{18}$ electrons, approximately.

This way, we define the units of one Coulomb and one Amper. Now, let define the electric tension between two points (charged particles), where the unit of tension is one *Volt*. The common definition is that one Volt is the electric potential between two points of a conducting wire,if an electric current of one Ampere dissipates one Watt of power. It can also be defined as the potential difference between two points that will impart one joule of energy per coulomb of charge that passes through it.

$$V = \frac{Jpul}{Coulomb} = \frac{kg \times m^2 \times s^2}{Amper \times s} \tag{1.1}$$

Now, we can define the *electric resistance* (R).

© The Author(s), under exclusive license to Springer Nature Switzerland AG 2023

S. Raczynski, *How Circuits Work*, Synthesis Lectures on Engineering, Science, and Technology, https://doi.org/10.1007/978-3-031-34934-8_1

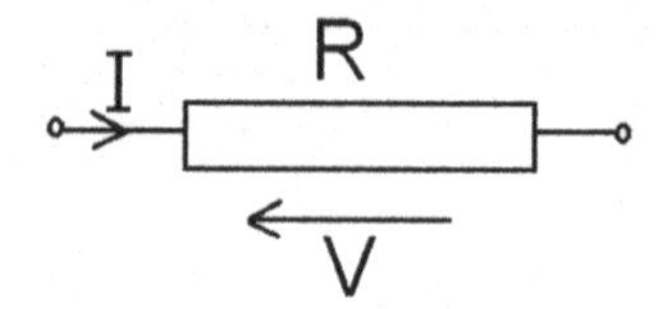


**Fig. 1.1** Current and voltage on a resistor

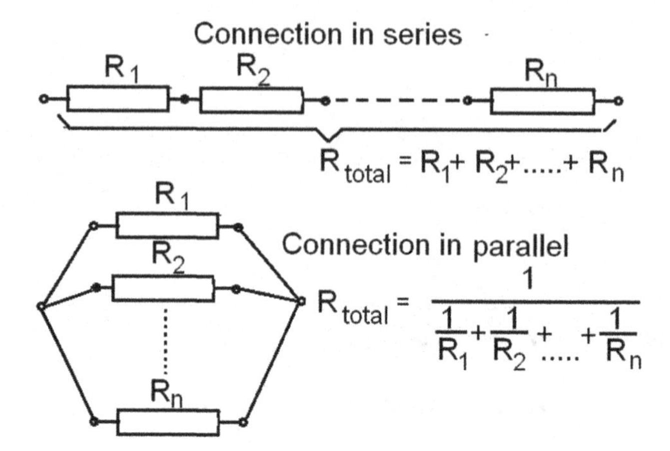


**Fig. 1.2** Resistance connections

Georg Simon Ohm (1789–1854) defined the relation between the tension (voltage) and the electric resistance, in his "Law of Ohm". Ohm's law states that the current through a conductor between two points is directly proportional to the voltage across the two points. The constant of proportionality is now called one *Ohm*, and usually denoted by $R$.

$$V = R \times I, \tag{1.2}$$

where $R$ is the electric resistance [Ohms], $V$ is the voltage over the resistance [Volts] and $I$ is the electric current [Ampers]. The unit of the voltage is called on *Volt*, after the Italian physicist Alessandro Antonio Volta (1745–1827). In the following, by an *ideal conductor* we mean a device (e.g.a wire) that has resistance equal to zero (Fig. 1.1).

The very elemental facts of circuits composed by two or more resistors (conductors with resistance R) are as shown in Fig. 1.2 (rectangles represent resistance).

### 1.1.2 Electric Power

The power is the measure of the rate of energy transferred, dissipated or received. The SI unit of power is the *watt*, one joule per second.

If there is a voltage $V$ and current $I$ on a device (for example a resistance), the corresponding power is equal to $P = VI$. Thus, for the resistance, we have

$$P = \frac{V^2}{R} = I^2 R$$

## 1.2 Circuits and Devices

By the *electric circuit* we understand a network of devices that contains one or more closed paths. The devices considered here are shown in figure. The electric resistance has been defined above, see Eq. (1.2). Other elements are as shown in Fig. 1.3. In this section, we consider some basic *passive* circuit elements. These are devices that do not amplify signals and can only store or dissipate energy.

The figure shows two different, commonly used symbols for the resistance.

**Capacitance**. This is a device that accumulates electric charge. If we apply a current $i(t)$ to the capacitance, as shown in Fig. 1.4, then the accumulated charge $Q$ and voltage $V$ on the capacitance are given by the Eq. (1.3):

$$Q(t) = \int_0^t i(\tau)d\tau, \quad V(t) = \frac{Q(t)}{C}. \tag{1.3}$$

Here, $Q$, $V$ and $i$ are functions of time. The constant $C$ is the size of the capacitance named Farad (F), after an English scientist Michael Faraday (1791–1867). We have one Farad of capacitance, if the charge of one Coulomb produces one Volt between the terminals of the device. The energy accumulated on the capacitance $C$ is equal to $CV^2/2$.

**Inductance**. The flow of electric current creates a magnetic field around the conductor, depending on the magnitude of the current. *Inductance* is defined as the ratio of the induced

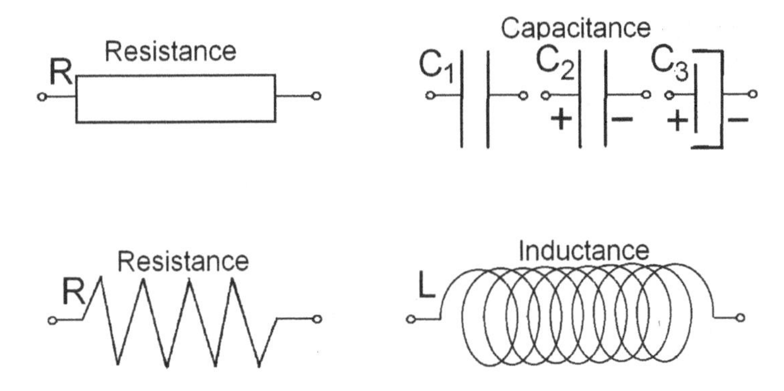


**Fig. 1.3** Symbols for RLC

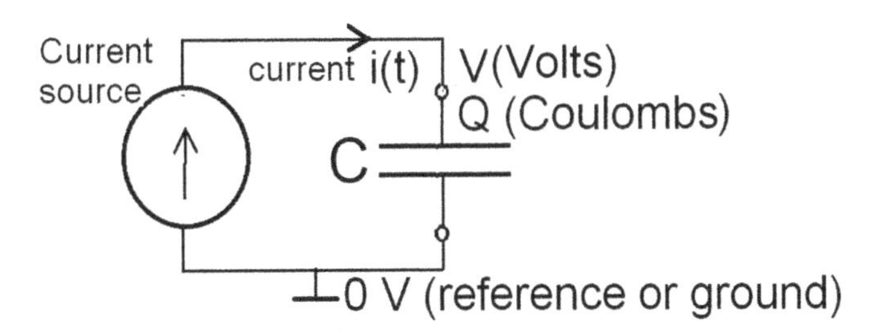


**Fig. 1.4** The capacitance

voltage to the rate of change of current causing it. The inductance is normally denoted as L in electric circuits. The equation that obeys the inductance is as follows.

$$V(t) = L\frac{di}{dt}, \tag{1.4}$$

where $L$ is the amount of inductance. The unit of inductance is one Henry (H), after an American scientist Joseph Henry, This is the amount of inductance that causes a voltage of one volt, when the current is changing by one Ampere per second. The inductance or *inductor*, typically consists of a coil or helix of wire.

These are the basic elements of electric circuits. Other devices like transformers, diodes, electronic tubes or transistors will be described in the following chapters.

## 1.3 Voltage and Current Sources

To use an electric device or circuit we must provide certain energy or electric power during a time interval. The sources of the power are mainly *voltage source* and *current source.*

Figure 1.5a shows the symbol of the ideal voltage source. *"Ideal"* means that this device has no internal resistance. It provides a fixed voltage between terminals a and b, independently on the circuit connected between a and b. Note that the **ideal voltage source cannot be connected to zero-resistance device** (ideal conductor or "short/circuit").

In Fig. 1.5b we can see an ideal current source. This device provides a given, fixed current $i$ that flows from it. Note that the **ideal current source cannot be connected to a circuit with infinite total resistance** (it cannot be "open circuit"). Part C of the figure shows the symbol of a controlled current source, where the amount of current $i$ is controlled by the current $i_0$, and $i = Ki_0$.

Figure 1.6 shows the real voltage and current sources. The real voltage source has an internal resistance R. So, if we connect a load between points a and b, the load voltage $V_1$ is less than the (nominal) source voltage $V_0$.

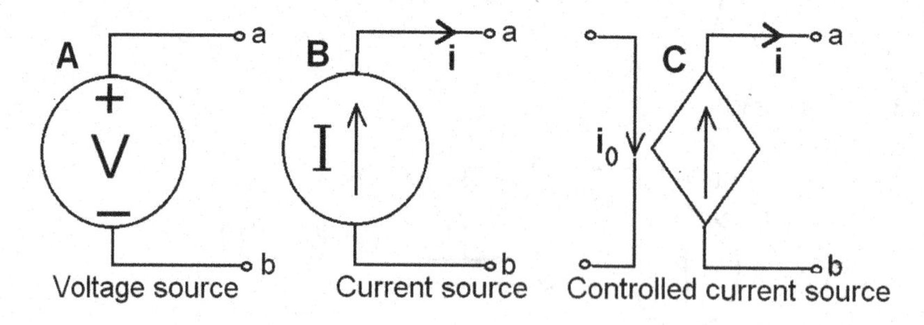


**Fig. 1.5** Ideal voltage and current sources

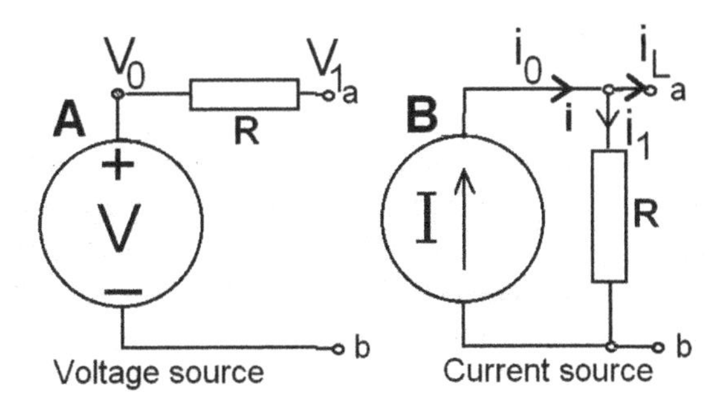


**Fig. 1.6** Real voltage and current sources

The real current source has a "leak" internal resistance. Consequently, the obtained current $i_L$ is less than the nominal current $i_0$. In practice, it is supposed that a good voltage source should have very low internal resistance, and the current source should have a big internal resistance. However, these resistances should not be forgotten while designing circuits. The *transformer* device will be discussed in Sect. 1.8.

## 1.4 Resolving Circuits

An electric circuit consists in a set of devises like resistors, sources or capacitances that form a red with one or more nodes and closed loops. Here, we do not discuss the circuit theory and analysis methods. Let only recall some facts that will be useful in the following.

### 1.4.1 Linearity

An electric circuit is *linear* if it obeys the *principle of superposition.* if we apply a signal $ax_1(t) + bx_2(t)$ to such circuit, then the response of the circuit (voltage and current in any part of the circuit) must be a linear combination of the response to $x_1(t)$ and $x_2(t)$, applied separately. So, the output signal F(*) should be as follows (Eq. 1.5):

$$F\,(ax_1(t) + bx_2(t)) = aF(x_1(t)) + bF(x_2(t)), \tag{1.5}$$

for any $a, b$ and time instant $t$. The rules discussed in the following sections are valid for linear circuits.

### 1.4.2 Voltage Divider

A simple and widely used circuit is a *voltage divider*.

If we apply the voltage $U$ between terminals a and b, then the voltage $V$ is equal to $R_2/(R_1 + R_2)$ (Fig. 1.7 part A). The same circuit with variable $R_1$ and $R_2$ is shown in part B of the figure. This version of the divider is known as the *Potenciometer*. The Potenciometer does not measure anything. It is widely used for audio volume control and for adjusting parameters of instrumentation devices.

### 1.4.3 Connection in Series and in Parallel

Useful formulas for the series and parallel connection of resistances are (see Fig. 1.8):

$$\begin{cases} R = R_1 + R_2 & \text{for connection in series} \\ R = \dfrac{R_1 R_2}{R_1 + R_2} & \text{for parallel connection} \end{cases} . \tag{1.6}$$

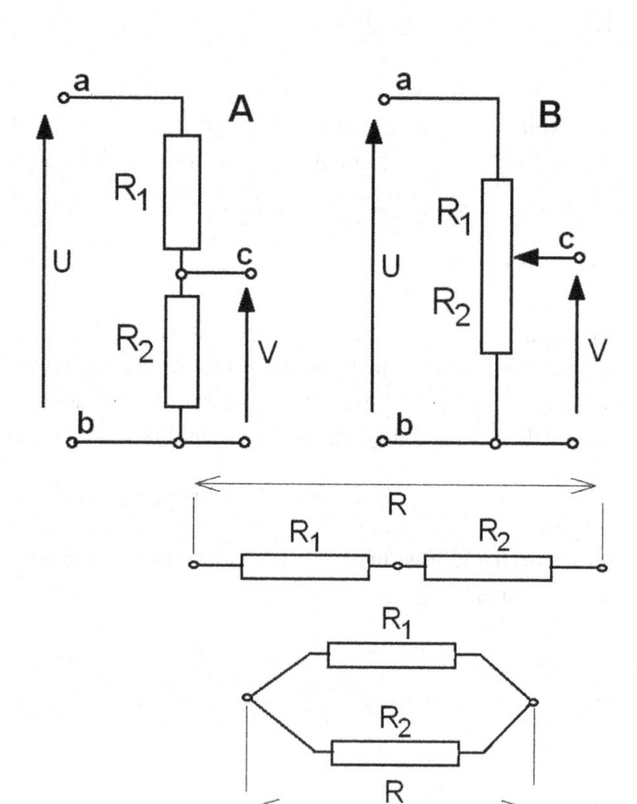


**Fig. 1.7** Voltage divider

**Fig. 1.8** Series and parallel connections of resistances

### 1.4.4 Loops and Nodes

Figure 1.1 shows the current and voltage produced on the resistor. Note that the voltage (potential difference between the resistor terminals) is opposite to the current direction, and is equal to $V = IR$.

The most elemental rules in circuit analysis are as follows:

* The sum of all currents at a node is equal to zero. Entering currents have positive sign and outgoing currents are negative.

* The sum of voltages over all devices in a closed loop is equal to zero.

These simple rules can be used to resolve circuits. Consider, for example, the circuit of Fig. 1.9. We have:

$$\begin{cases} i_0 - i_1 - i_3 = 0, & \text{A (node a)} \\ i_1 - i_2 - i_4 = 0, & \text{B (node b)} \\ i_3 + i_2 - i_5 = 0, & \text{C (node c)} \end{cases} \tag{1.7}$$

From the loops $V - R_1 - R_4$, $R_2 - R_5 - R_4$ and $R_3 - R_2 - R_1$ we obtain:

$$\begin{cases} V - i_1 R_1 - i_4 R_4 = 0, & \text{D} \\ i_4 R_4 - i_2 R_2 - i_5 R_5 = 0, & \text{E} \\ i_3 R_3 - i_2 R_2 - i_1 R_1 = 0, & \text{F} \end{cases} \tag{1.8}$$

The above six equations (A, B, C, D, E and F) allow us to calculate six unknown variables $i_0, i_1, i_2, \ldots, i_5$ that resolves our circuit.

Now, consider a circuit composed by several voltage sources and a red of resistors. Select two nodes A and B that are not both terminals of the same source. Using the above rules, we can resolve the circuit and calculate the voltage $V_A$ and $V_B$ on the nodes, respectively. Then, connect the nodes with a resistor $R_x$ and resolve the circuit again to get the current flowing through $R_x$. Now, pass with $R_x$ to zero, $R_x \to 0$. Denote the limit current on $R_x$ by $i_x$.

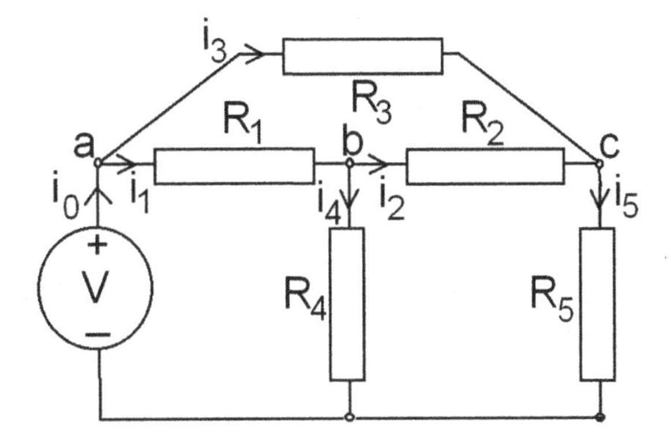


**Fig. 1.9** Example circuit

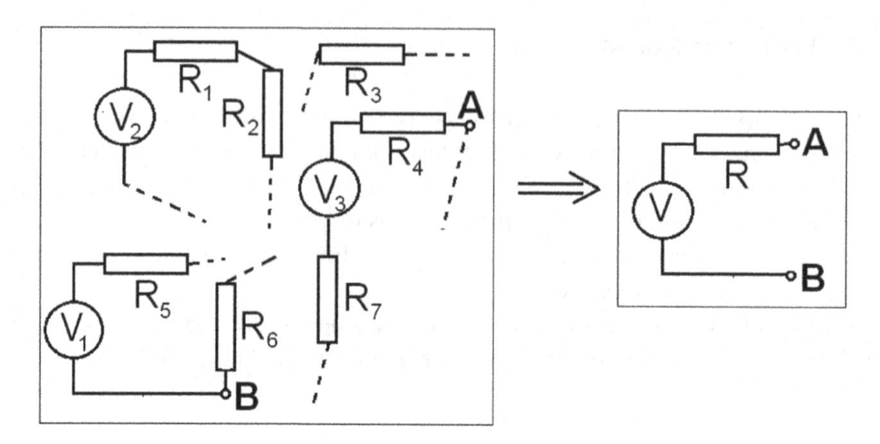


**Fig. 1.10** The theorem of Thevenin

Our circuit, viewed from the nodes A and B can be replaced by a voltage source $V = V_A - V_B$ connected in series with the resistance $R = V/i_x$ (see Fig. 1.10). This equivalence is a case of the *theorem of Thevenin* that allows us to simplify circuit analysis [2].

Other useful rule is given by the *Norton's theorem*, where the simplified equivalent circuit consists of a current source with a resistor connected in parallel.

We will not provide more facts from the circuit theory because this is not the main topic of this book. The above remarks are sufficient to resolve the circuits discussed in the following.

## 1.5 Direct and Alternate Current

### 1.5.1 DC and AC Supply, Effective Voltage

By *direct current* we mean an electric current and/or voltage that does not change in time. The power produced by such current has been mentioned in Sect. 1.1.2.

Now, suppose that the current or voltage are functions of time. As for the power supply, those are the AC sources, that provide this kind of electricity. It is supposed that an AC source produces the voltage/current that is sinusoidal, described by the Eq. 1.9

$$v(t) = Asin(2\pi ft + \varphi) \tag{1.9}$$

where $f$ is the frequency [Hz], $t$ is the time, and $\varphi$ is the fphase-shift. By the *angular frequency* we mean $w = 2\pi f$, [rad/s].

The amplitude $A$ in Eq. (1.9) is the maximal absolute value reached by the voltage $v(t)$. If we apply the voltage $v(t)$ to a resistance, the average power produced over one period

T is equal to $A^2/(R \times sqrt(2)) = V_e^2/R$, where $V_e$ is called the *effective voltage*. For AC power supply, the source voltage is normally given as the effective voltage. So, if we have a source of 127 V AC, then the maximal instantaneous voltage $v(t)$ is equal to $127\sqrt(2)$, approximately 179.6 V. It is important to remember that the above relations are valid only for the sinusoidal form of voltage and current.

There are some expressions and kinds of the electric power. Consider the sinusoidal voltage provided by a supply source. The Volt-Amper (VA) power is just $P = Vi$ where $V$ is the effective voltage, and $i$ is the current. The *instantaneous power* is $P(t) = V(t)i(t)$. The *real power* is $P = Vi\cos(\varphi)$, where $\varphi$ is the phase-shift between voltage and current. For example, it the load is an ideal inductor, then the current is delayed (in phase) by 90° with respect to the voltage. So, the power produced on the inductor is equal to zero, though both the voltage and current may be great. The *imaginary power* is equal to $P = Visin(\varphi)$.

## 1.6 Signals

In the following, we discuss circuits that are receiving and processing signals given as functions of time, see Fig. 1.11.

A signal may be given as a changes of voltage as well as of the current. Two main components of signals are the *constant or DC* component and the *variable or AC* component. The constant component is supposed to be constant or to change very slowly. The variable component reveals fast changes around the DC level, as shown in Fig. 1.11.

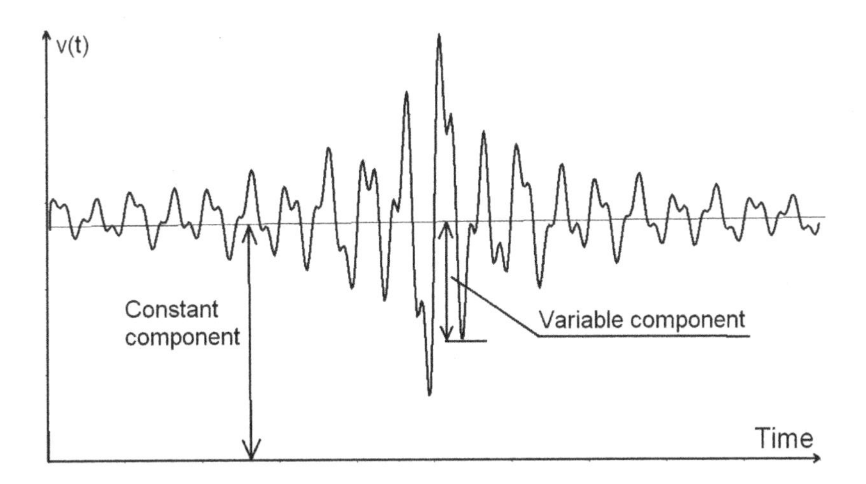


**Fig. 1.11** Constant (DC) and variable (AC) components of a signal v(t)

A simple sinusoidal signal may be defined as $v(t) = A\,sin(wt + \varphi)$, where $w$ is the angular frequency, $t$ is the time and $\varphi$ is the phase shift. Other signal may have the form of a rectangular or triangular waves. The more complicated signals are given as a superposition of many different frequencies.

### 1.6.1 Complex Numbers

It is supposed that the reader is acquainted with the concept of *complex numbers*. Recall only the main facts.

The unit of *imaginary number* is dented by $j$. In the complex numbers theory the letter $i$ is commonly used. However, in electrical engineering, we rather use $j$ to avoid confusion with the amount of electrical current, normally denoted as $i$. A *complex number* has the form $a + jb$.

We have

$$\begin{cases} j^2 = -1 & A \\ 1/j = -j & B \\ |a + jb| = \sqrt{(a^2 + b^2)} & C \text{ (Absolute value)} \\ \angle(a + jb) = arctan(b/a) & D \text{ (Angle)} \\ e^{j\varphi} = cos(\varphi) + j sin(\varphi) & E \text{ (Euler's representation)} \end{cases} \tag{1.10}$$

For two complex numbers $x$ and $y$ we have $xy = z$ where $|z| = |x| \times |y|$ and $\angle z = \angle x + \angle y$ (Fig. 1.12).

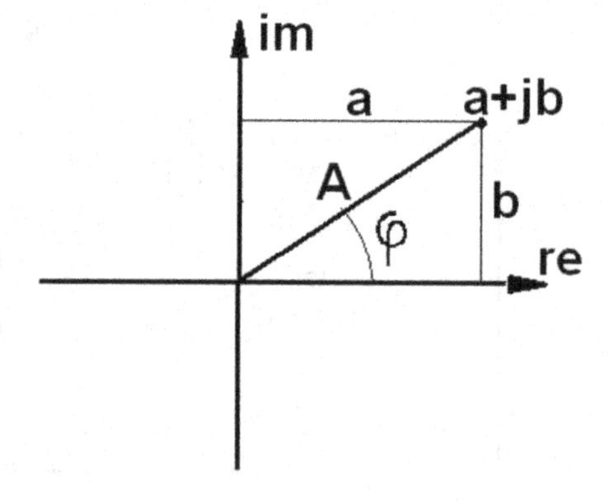


**Fig. 1.12** The complex number

### 1.6.2 Sinusoidal Signal

The AC component of the sinusoidal voltage signal is as follows.

$$v(t) = A\,sin(wt + \varphi), \tag{1.11}$$

where $A$ is the signal amplitude, $w = 2\varphi f$ is the angular frequency [rad/s], $f$ is the frequency in Hz, $t$ is the time, and $\varphi$ is the phase-shift.

In circuit analysis we will use the complex form of the sinusoidal signal, as shown in Eq. (1.12).

$$v(t) = A\ (cos(wt + \varphi) + j sin(wt + \varphi)) = Ae^{j(wt+\varphi)} \tag{1.12}$$

### 1.6.3 Spectrum of the Signal

The spectral analysis and the theory of the Fourier transform are not topics of this book. Let us only recall that a real, arbitrary signal $v(t)$ can be treated as a superposition of finite or infinite number of sinusoidal components. The transformation of Fourier is used to analyze the signal given in form of function of time, and to calculate the included frequencies (consult [3]).

By the *signal spectrum* we mean the description of a signal as a function of frequency. It represents the signal as the superposition of the frequencies that it contains. The spectrum is shown graphically as a plot where the independent variable is the frequency, and the dependent variable is the corresponding amplitude.

Figure 1.13 shows an example of the spectrum of an acoustic signal.

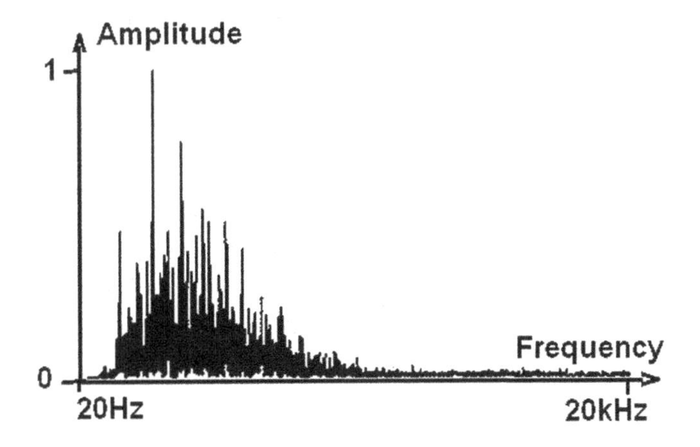


**Fig. 1.13** A signal spectrum

## 1.7 The Impedance

Looking at the equations for the capacitance and inductance Eqs. (1.3) and (1.4) we can deduce that for sinusoidal form of the current on the capacitance, the voltage is delayed by 90° on the complex plane with respect to the current. For the inductance, the current is delayed in phase by 90°. Remember that phase shift by 90° and −90° correspond to multiplication by $j$ and $-j$, respectively. As the consequence of (1.3) and (1.4) we have:

$$\begin{cases} Z_C(w) = \dfrac{1}{Cjw} & \text{for capacitance} \\ Z_L(w) = Ljw, & \text{for inductance} \end{cases} \tag{1.13}$$

where $Z(w)$ is the impedance. The expressions for current and voltage ($I$ and $V$)are as follows.

$$\begin{cases} V = \dfrac{I}{Cjw} & \text{for the capacitance} \\ V = ILjw & \text{for the inductance} \end{cases} \tag{1.14}$$

Note that for the capacitance the current is the cause, and the voltage is the result. For the inductance with applied voltage, the current is the result.

The impedance can be used in similar way as the resistance when calculating impedances of combinations of the R, L and C devices. For example, the connection in series of a capacitance C and an inductance L has the total impedance as follows.

$$Z_{LC}(W) = Ljw + \frac{1}{Cjw}$$

Note that if $w = 1/\sqrt{LC}$, then the above impedance is equal to zero.

On the other hand, consider the connection of inductance and capacitance in parallel.

Looking at the impedances given by Eq. (1.13), we can see that when the frequency $w$ grows, then the impedance of the inductor also grows, and the impedance of the capacitance becomes small. This can be used while looking for simplified equivalent circuits. Consider a simplified circuit composed by resistances and capacitances, for the DC signal component. In such circuit, we can ignore the capacitances, considering them as open connections (infinite impedance). For the AC component we can replace the capacitors by short-circuits (suppose $Cw >> 1$).

## 1.8 The Transformer

Figure 1.14 shows the symbol of the *transformer*. Transformers are used to convert the alternate (AC) voltage applied to the primary winding, into the voltage on the secondary winding. This device consists of two windings if $n_1$ and $n_2$ turns respectively, over the common ferromagnetic core. In the core a magnetic flow is induced, common for the two windings.

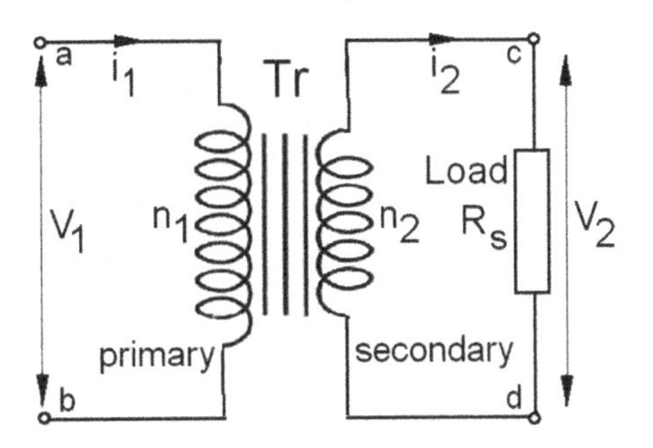


**Fig. 1.14** Transformer

For an *ideal transformer* we have the following relations:

$$\begin{cases} V_2 = \frac{n_2}{n_1} V_1 \\ i_2 = \frac{n_1}{n_2} i_1 \\ R_p = \left(\frac{n_1}{n_2}\right)^2 R_s, \end{cases} \tag{1.15}$$

where $R_p$ is the load resistance viewed from primary terminals a, b ($R_s$ transformed to the primary side).

In a real transformer we have some (unwanted) impedances and currents that may modify the relations (1.15). The following facts have to be taken into account.

* The two windings are made of wires that have finite resistance.

* Not all magnetic flow produced by the primary winding is captured by the secondary winding. This results in an additional inductance of the primary winding.

* if there is no load the primary current is not equal to zero. Some "no load" current is needed to magnetize the core.

* For excessive load and high currents, the histaresis of the core makes the transformer non-linear and the considerable loses of power occur.

Figure 1.15 shows the equivalent circuit of a real transformer that can be used in circuit analysis. The "Tr" element of the circuit is the ideal transformer.

Where,

Tr—ideal transformer

$Z_1 = R_1 + jL_1$ Primary winding impedance.

$Z_2 = R_2 + jL_2$ Secondary winding impedance.

$Z_2$ Secondary winding impedance.

$i_0$ No-load current.

$i_t$ = Primary current of ideal transformer.

$i_\mu$ Magnetizing Component,

$i_w$ Working Component,

No-load current is a vector summation of working component $i_w$ and magnetizing component $i_\mu$.

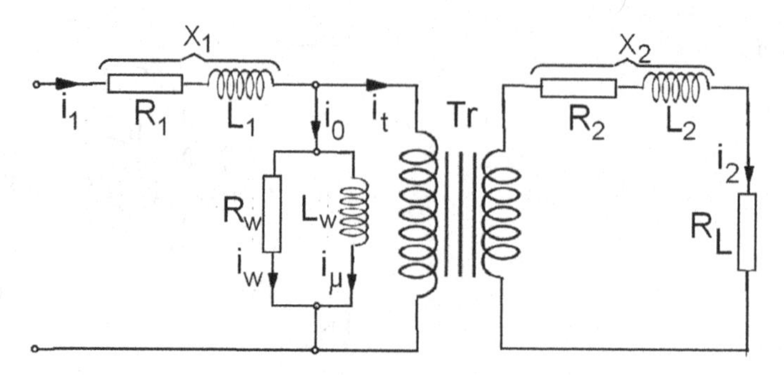


**Fig. 1.15** Transformer equivalent circuit

In a preliminary design of circuits with transformers, we can use the ideal transformer. However, if some more exact solution is required, then we should use the equivalent transformer circuit. Note that in the equivalent circuit, inductances appear in series with the currents. Consequently, the high frequencies may not pass correctly to the load. The resistances in the equivalent circuit imply energy loses that elevate the transformer temperature.

## 1.9 Diodes, Triodes, Transistors and Other Devices

Now, let see some non-linear and active devices. By an *active device* we understand those that can amplify current, voltage and the electric power, and can be controlled by external signals.

### 1.9.1 Electronic Tubes

The *electron tube*, known also as a *valve, vacuum tube* or *tube* is a tube made of glass with vacuum inside. In the tube there are two or more *electrodes*. These devices are mentioned here for historical reasons. The use of electronic tubes is recently very low, though some instruments based on tubes are still in operation.

### 1.9.2 Diode Tube

In 1904, John Ambrose Fleming invented the *electronic diode*. He observed that if one of the electrodes is heated, and a voltage difference is applied to the device, then a flow of electrons appears from the heated electrode (cathode), and the other electrode (anode). The

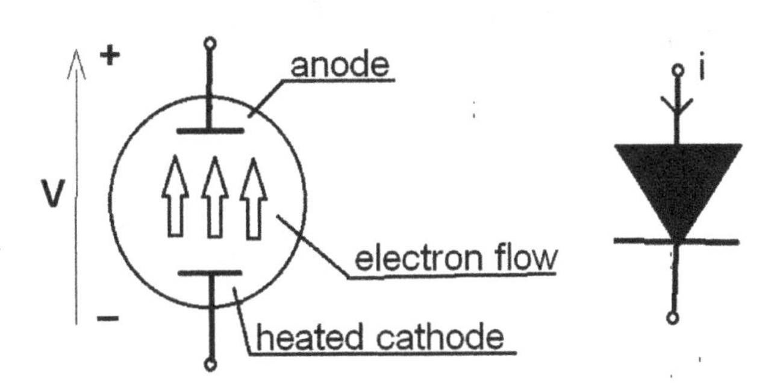


**Fig. 1.16** A diode tube

anode must have positive polarization. If the polarization is opposite, no electrons flow inside the tube. Figure 1.16 shows the scheme and a symbol of the diode tube.

Diodes are used to rectify an AC signals and power supply. If the applied voltage has a sinusoidal form with DC component equal to zero, then the DC component of the diode current is different from zero.

### 1.9.3 The Triode

The current inside the tube can be controlled by the voltage of a grid made of thin wire, located between the cathode and anode. This converts the tube in an amplifying devise. The DC voltage between the grid and the cathode is negative. Small fluctuations of the grid voltage result in considerably big changes in the cathode-anode current. The great advantage of the triode is that the grid current is very small (some microampers) compared to the cathode-anode current (Fig. 1.17). This allows to construct a good quality amplifiers. Triodes are used mainly for pre-amplification circuits. One of the disadvantages of the triode is a small, but relevant capacitance that always exists between the anode and the grid. This capacitance results in a smaller gain for high frequency signals. To avoid such unwanted capacitance, an additional grid is added between the firs grid (the controls the current) and the anode (a screen grid). The voltage if the additional grid is fixed. This grid separates grid one from the anode. Such tube is called *tetrode*.

In the *pentode tube* the third grid is added, called the *suppressor* grid. This grid prevents the secondary emission of electrons from the anode to reach the screen grid. Penthodes are frequently used for the power amplification stages of the amplifier.

As mentioned before, we recall these properties of electronic vacuum tubes for historical reasons. We will not follow with tube applications, and dedicate the rest of this book to semiconductor devices.

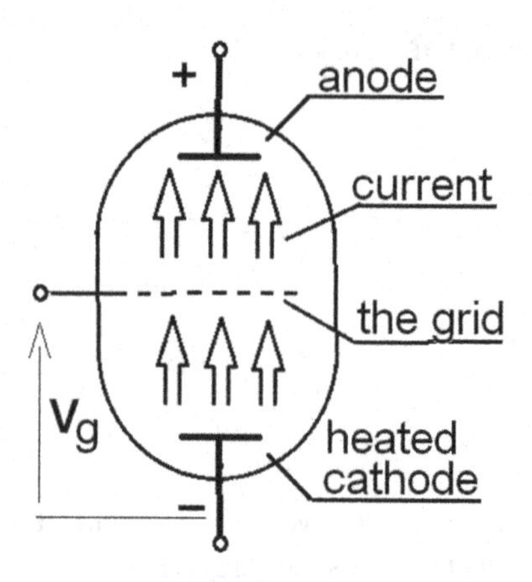


**Fig. 1.17** A triode

## 1.10 Semiconductor Diodes and Transistors

The first semiconductor devices have been based on germanium. Other semiconductors like silicon, gallium arsenide can be used. Recently the silicon devices dominate.

The crystal lattice of the semiconductor may occasionally have "holes" where there is a lack of an electron. The hole becomes a spot with positive electric charge. The semiconductor with holes is called "p-type" (positive charged). The region with free electrons is of type "n". Holes can move over the lattice. This, together with free electrons, results in electric conduction. Electrical current may pass through the junction p-n only in one direction that occurs in semiconductor diode.

The p- and n-type regions are formed by ion implantation, diffusion of dopants, or by epitaxy when the crystal lattice grows.

### 1.10.1 Diodes

A German physicist Ferdinand Braun in 1874 discovered a *semiconductor diode*. The diode has a p-n junction in silicon, gallium arsenide or germanium, connected to two electrical terminals. The device has a very low resistance in the diode's *forward direction*, and very low in the opposite, *reverse direction*. Diodes are used to rectify AC current in the DC current power supply, to convert radio signals in audio-frequency oscillations, and in many other applications. Figure 1.18 shows a typical diode characteristics. Observe that in the forward direction, the diode needs a small forward voltage (about 0.6 V) to conduce. In the reverse direction, there is a considerably great *breakdown voltage*, when the diode may be destroyed.

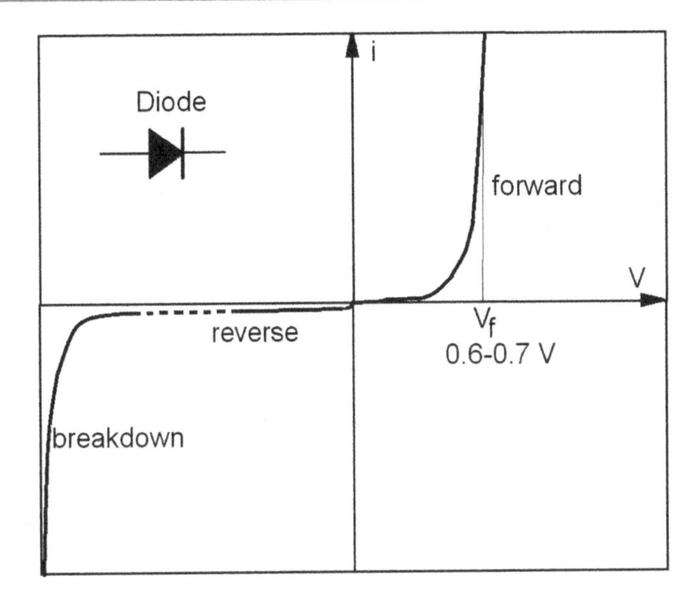


**Fig. 1.18** Semiconductor diode

The *ideal diode* is a device that has resistance zero in forward direction, and infinite in the other. In the simplified circuits, to the ideal diode is frequently added the forward voltage as an additional voltage source.

A **Zener diode** is similar to a conventional diode, except the reduced breakdown voltage. This voltage is well defined and stable. This permits to use the Zener diode as a reference voltage source. Figure 1.19 depicts the symbol of the Zener diode and an application.

There are many other types of diodes, developed for specific applications. Shortly speaking, there are the following diode types (see Fig. 1.20).

*Light emitting diode (LED)* emits light when current flows through it. Discovered in early 1960s, the recently used LEDs can emit a considerable amount of light.

*Photodiode* is a type of P-N junction diode that converts the light energy into electrical current. Its operation is opposite to that of an LED.

*Laser diode* is similar to LED. It converts electrical energy into light energy. However, unlike LED, Laser diode produces coherent light.

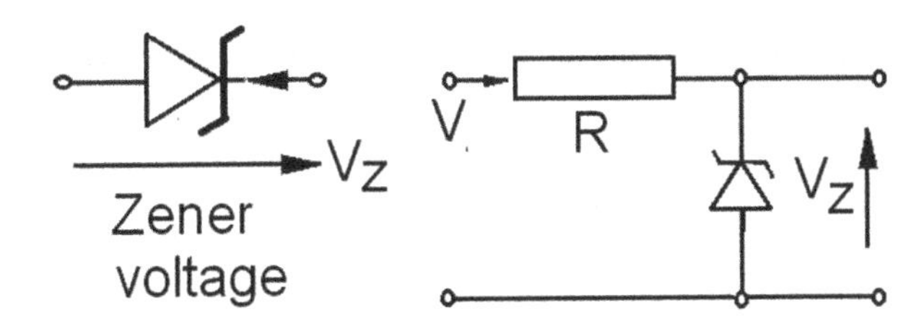


**Fig. 1.19** Zener diode

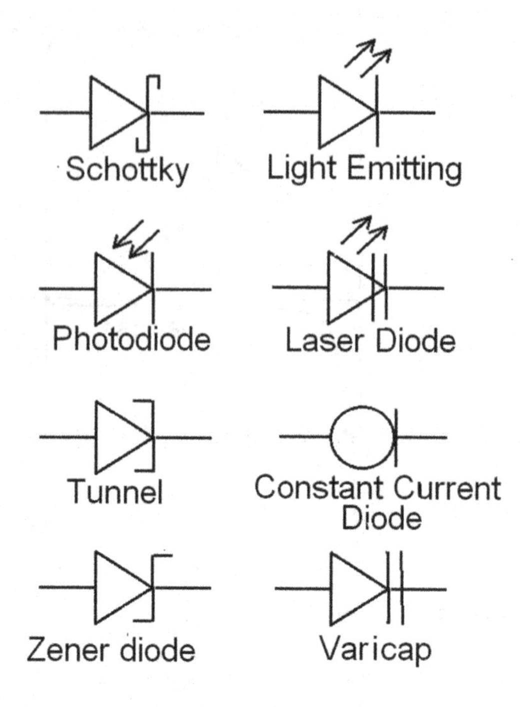


**Fig. 1.20** Diode types

*Tunnel diode* invented by Leo Esaki in 1958, also known as Esaki diode, works on the principle of the *tunneling effect*. Due to heavy doping concentration, the junction barrier becomes very thin. This allows the electron to easily escape through the barrier (tunneling effect). In the characteristic of this diode there is a region of negative resistance. The diode has various application, mainly as a fast switching device.

*Schottky diode* that has small junction between an N-type semiconductor and a metal, and practically no capacitance of the junction. It has very low forward voltage drop and fast switching. The Schottky diode switching speed is very fast.

*Avalanche diode* is a P-N junction diode that is specifically designed to operate in the avalanche breakdown region. It works in similar way as the Zener diode, but permits a heavy current flow in reverse direction.

*Constant current diode* of *current limiting diode* is made from JFET. It regulates the current flow through it up to a fixed level.

*Varicap diode* is variable capacitance device. The internal capacitance of the diode is controlled by the voltage applied as reverse bias across the varicap. This device can be used to control the frequency of an LC oscillator, mainly to get the *frequency modulated* (FM) signals.

### 1.10.2 Ideal Versus Real Components

In previous sections, there are several references to the "ideal" components. In many cases, and in a preliminary design, we often assume that the components are ideal. For example, a wire with no resistance, capacitance and inductance, a capacitor with no internal resistance or inductance, an inductor without capacitance or resistance, etc. Such approximation work to some extent. However, we should be careful assuming ideal devices. Let us see, for example, the "two capacitor puzzle" (Fig. 1.21).

Assume the ideal devices: the wires without resistance, capacitance or inductance, capacitors with no internal resistance, capacitance or inductance, $C_1 = C_2$ both equal to $C$. In the initial condition, capacitor $C_1$ has voltage $V_0$, and the voltage of $C_2$ is equal to zero, switch S open. The total energy of the circuit is $E_0 = CV_0^2/2$.

Now, we close the switch. Logically, after closing S, the electric charge distributes between two capacitors, so each of them has the voltage $V = 0.5\ V_0$. Let us calculate the total circuit energy:

$$E = 2[C(V_0/2)^2/2] = E_0/2$$

The question is: Where the half of the initial energy went?

The answer is as follows. We have assumed ideal wires. Let us allow the wires to have a very small resistance $R$. Passing with R to zero, we should obtain our idealized circuit. If $R > 0$, then the current i obeys the equation:

$$i(t) = V_0 e^{-0.5RCt}.$$

The instant power produced on $R$ is $P(t) = i(t)R$. Integrating $P(t)$ over the interval $[0, \infty]$, we can see that the energy dissipated on $R$ is equal to half of the initial energy of the circuit, so, the energy balance is OK. Moreover, this energy does **not** depend on $R$ (we left the proof to the reader). Now, pass with $R$ to zero. It should be expected that, in the limit, we obtain our ideal circuit. However, this is not true. In the limit, the dissipated energy is still equal to $0.5\ E_0$. So, what is wrong with our ideal circuit? The answer is: **The ideal original model is invalid**. The sequence of models with $R \to 0$ does not converge to the idealized model, in terms of the energy balance. The validity of models is an important

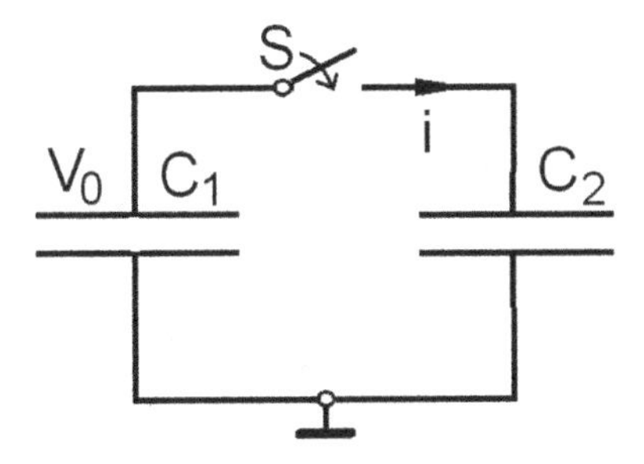


**Fig. 1.21** Two capacitors and a switch

issue in modeling and simulation. Treating with circuits, the model validity can hardy being checked. We assume that the differential equations of circuit variables are valid, and we use them to analyze circuits. This, in most cases, is correct, but we must remember that not all models of idealized circuits are valid, and may lead to wrong results.

### 1.10.3 The Transistor

The *transistor* was invented by Julius Edgar Lilienfeld in 1925. Further research done by Oskar Heil, Robert Pohl y Rudolf Hilsch contributed to the device called now the *transistor*. It consists in two junctions between p and n type of semiconductor (see Fig. 1.22). The terminals are called *emitter, base* and *collector*, marked with letters E, B and C, respectively. Figure 1.23 shows the symbols of the two types of transistors. This type of transistor is often referred to as BJT (bipolar junction transistor). The transistor current $i_c$ is controlled by the current of the base $i_b$. A very small base current (some microampers) is needed to control a considerable big current of the collector (up to several ampers in big, power transistors). Note that $i_c = i_e + i_b$. In the NPN silicon transistor, the potential difference between base and emitter, in normal operation, oscillated between 0.5 and 0.7 V. In simplified equivalent

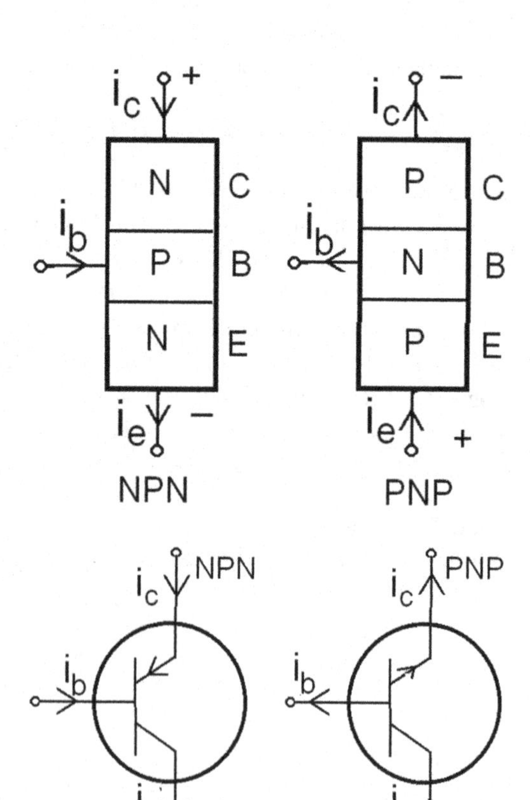


**Fig. 1.22** NPN and PNP transistors

**Fig. 1.23** Symbols of NPN and PNP transistors

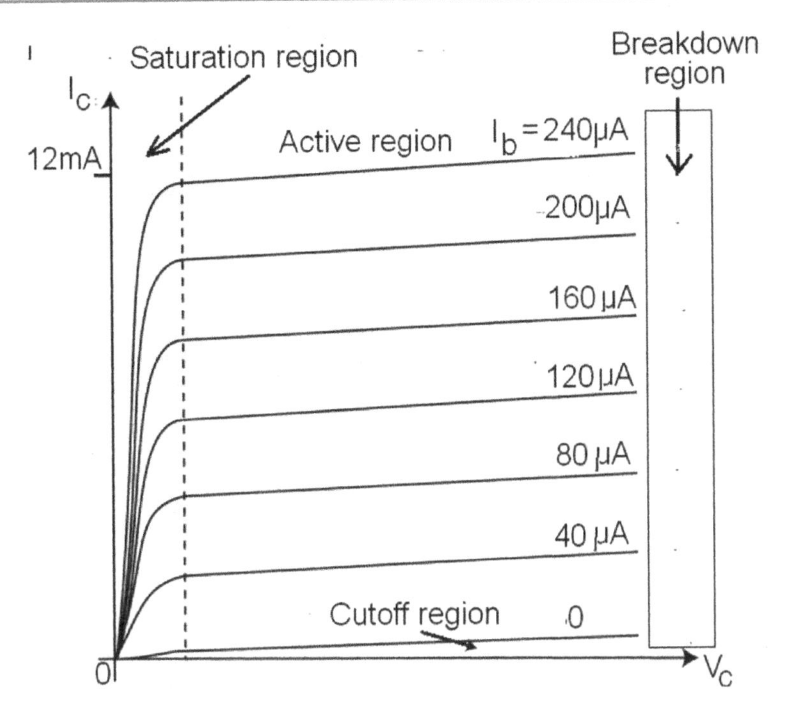


**Fig. 1.24** The $V_c - I_c$ characteristic of a BJT transistor

circuits it ca be assumed to be 0.6 V, approximately. The main parameters of a BJT transistor are: the nominal collector current, and the parameters $\alpha$ and $\beta$, defined by Eq. (1.16).

$$\begin{cases} \alpha = \dfrac{i_c}{i_e} & \beta = \dfrac{i_c}{i_b} \\ \alpha = \dfrac{\beta}{1+\beta} & \beta = \dfrac{\alpha}{1-\alpha} \end{cases} \tag{1.16}$$

Figure 1.24 shows the $V_c - I_c$ characteristics of a BJT low-current transistor, $\beta = 50$. The curves correspond to different base currents. Note that the base current is expressed in micro-Amperes, and the collector current in milli-Amperes. The "active region" is where the transistor should operate. Observe that for a constant base current, the characteristics are nearly horizontal. This means that the corresponding collector current is approximately constant. This property is used in simplified equivalent circuits, where we can assume that the transistor is a current source, for the variable component of the signal.

In the *field effect* transistors (FET or JFET), the current is controlled by the voltage of the "gate terminal" instead of base current. This makes the input resistance (base-emitter) practically infinite, and the base current can be neglected. The transistor has three terminals: source, gate, and drain, see Fig. 1.25.

Fig. 1.25 Field effect transistor

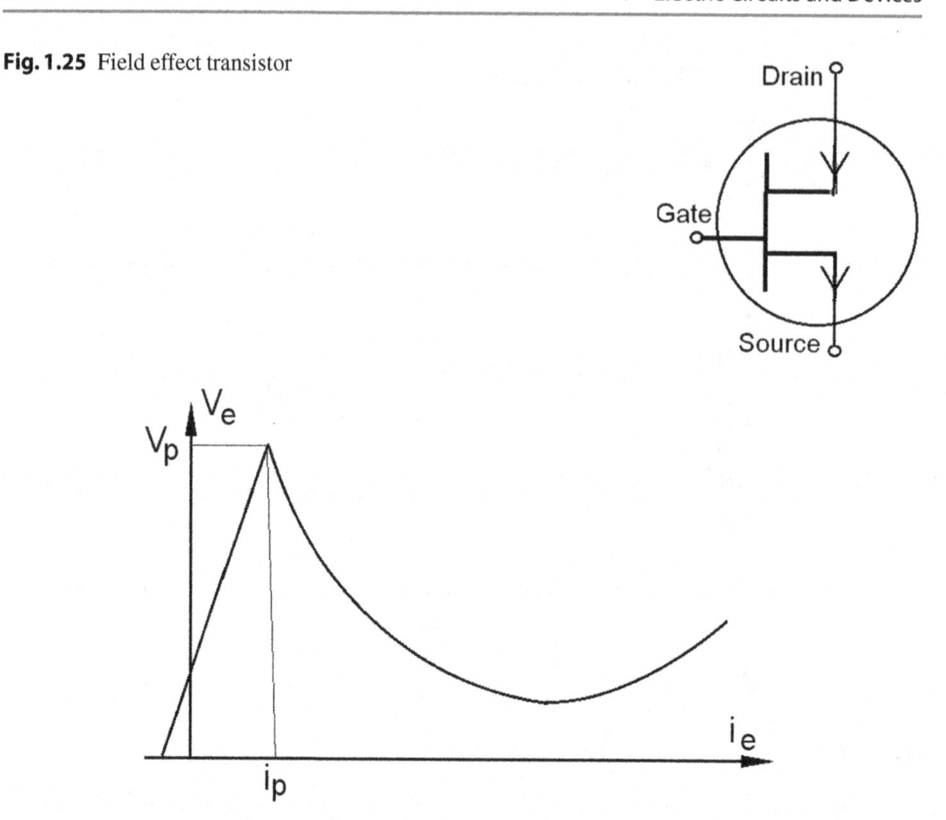


Fig. 1.26 Emitter current characteristic of UJT transistor

## 1.10.4 Unijunction Transistor UJT

This transistor has only one junction. It is used as a switching device.

UJT has three terminals: the emitter E, and the bases B1 and B2. The base B2 is connected to the supply voltage through a resistor, and base B1 by another resistance to the ground. While increasing the emitter-base one voltage and reaching certain level marked as VP, the emitter-B1 resistance becomes negative, and the emitter current rapidly grows, as shown in Fig. 1.26. This phenomenon is used mainly in relaxation oscillators and similar circuits where a switching device is needed.

## 1.10.5 CMOS Transistor

In early 1960 the Fairchild Company developed a transistor CMOS. It is based on metal-oxide semiconductor field-effect transistor (MOSFET) technology. CMOS stands for Complementary Metal Oxide Semiconductor. The field effect is also implemented in the *metal-oxide-semiconductor field-effect transistor* (MOSFET). Transistors of this type are similar

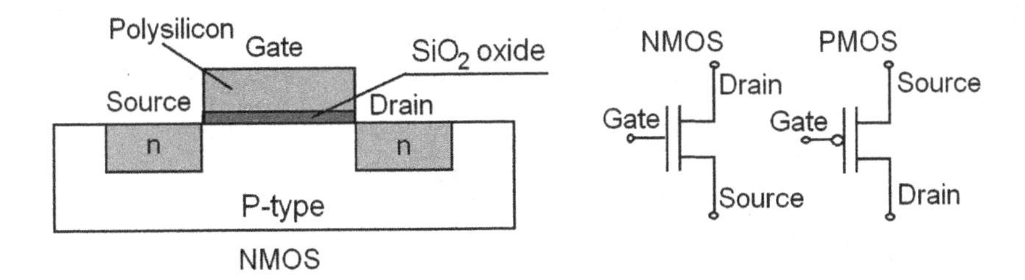


**Fig. 1.27** The CMOS

to JFET. They are easier to produce and support higher voltages. There are two basic types of this device: NMOS and PMOS (Fig. 1.27).

In NMOS, a p-type substrate is used which has n-type source and drain diffused on it. Here, the majority carriers are electrons. When a high voltage is applied to the gate, the NMOS conducts and when low voltage is applied to the gate, it does not conduct. NMOS is faster than PMOS, as the carriers are electrons that travel twice as fast as the holes.

P-MOS consists of P-type Source and Drain diffused on an N-type substrate. Holes are the majority carriers. When a high voltage is applied to the gate, the PMOS does not conduct and when a low voltage is applied, it conducts. These devices are more immune to noise than NMOS devices.

CMOS are mainly used as switching devices in logic gates and logical circuits. However, some application to the linear signal processing can be found.

### 1.10.6 The Thyristor

Consider a semiconductor device with three junctions, between four layers P-N-P-N, as shown in Fig. 1.28. The device can be interpreted as two transistors: PNP and NPN, interconnected as shown in the figure. The base of the PNP is, the same time, the collector of NPN. The emitter of the PNP is the gate of NPN. The thyristor has two possible states: it may be open (conducting) or closed (zero current anode-cathode).

First, suppose that we apply a positive pulse to the gate terminal. Then, the transistor NPN opens, and its collector current grows. This is, at the same time, the current of the base of PNP, so the PNP opens also. The thyristor can remain in this, stable state. Now, we apply a negative pulse to the gate. The NPN no longer conducts, so, the base current of PNP disappears. This means that the PNP also closes. This is the second, also stable state of the thyristor that remains closed, with null anode-cathode current.

Thyristors are used in power supply devises and control circuits, controlling the supplied power. The application circuits will be discussed in Chap. 5.

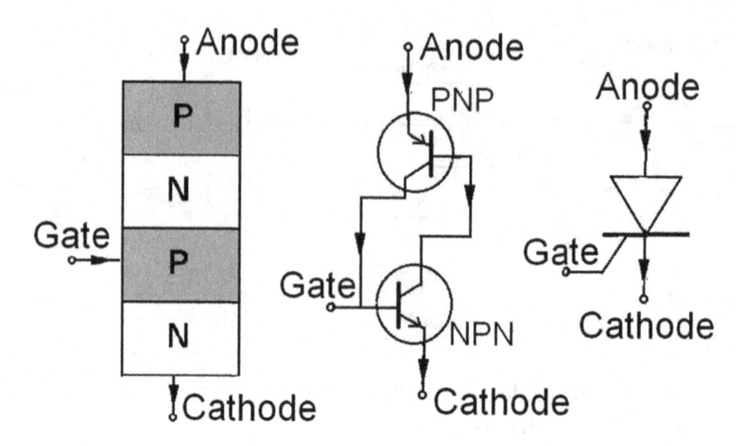


**Fig. 1.28** The thyristor

# Voltage and Power Amplifiers

# 2

## 2.1 Basic Voltage Amplifier

Let start with the simplest voltage amplifier based on a BJT NPN transistor. Figure 2.1 depicts the amplifier scheme.

The part B of the figure shows the equivalent simplified circuit for the variable component (AC). Note that for AC, the node VCC is the same as the ground G (AC component zero). So, the collector resistance $R_c$ and the load $R_L$ are connected in parallel. The transistor is replaced by a controlled current source. $R_{be}$ is the resistance base-emitter. This is a simplified circuit that can be used for the preliminary design. The resulting circuit should work satisfactorily. However, for the more exact design we soul characteristics of the transistor. It is recommended make a preliminary calculations as below, and then check and adjust, if necessary, the circuit parameters using a circuit simulation software like SPICE or CircuitMaker.

We want to obtain an amplified output voltage $v_{out} = K v_{in}$ on a load resistance $R_L$, $K$ being the voltage gain of the amplifier. In the part A of the figure we can see the electrical scheme of the amplifier. The input signal $v_i n$ is applied to the base of the transistor. The capacitance $C_1$ inhibits the DC component of the input to pass to the base. Now, we must make the collector-emitter current flow. Suppose that we have a voltage supply VCC, for example DC 15 V. We cannot apply this voltage to the collector because this would fix the collector AC component. So, we must use a resistor $R_c$ between VCC and the collector. To get the maximal possible range of oscillations at the collector C, the DC voltage at point C should be approximately equal to VCC/2. To calculate $R_c$, look at the parameters of the transistor. The manufacturer always provides the data about the optimal operation point that includes the normal current $i_c$, say, $i_{co}$. So, our resistor $R_c$ must be equal to $VCC/(2i_{co})$. Now, use the parameter $\beta$ of the transistor. If the collector current is equal to $i_{co}$, then the base DC current must be $i_b = i_{co}/\beta$. Having $i_b$, we calculate $R_b = (VCC - 0.6)/i_{co}$. Finally,

© The Author(s), under exclusive license to Springer Nature Switzerland AG 2023

S. Raczynski, *How Circuits Work*, Synthesis Lectures on Engineering, Science, and Technology, https://doi.org/10.1007/978-3-031-34934-8_2

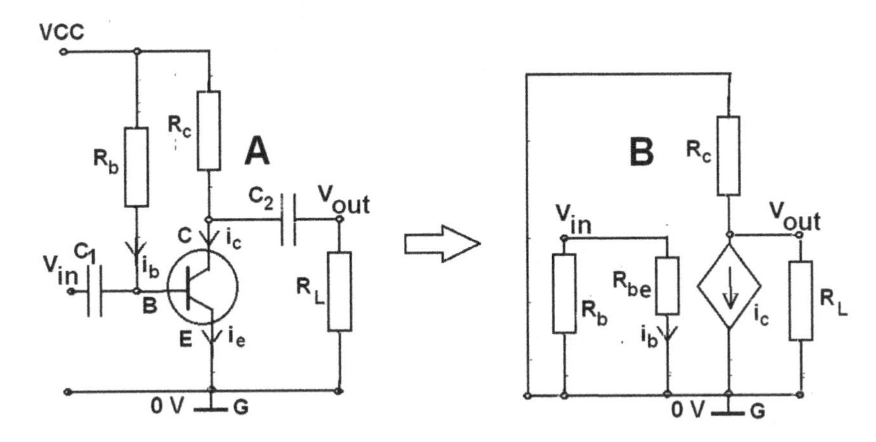


**Fig. 2.1** Simple voltage amplifier

we calculate the gain of the amplifier, $v_{out}/v_{in}$. Let denote by $R_{cL}$ the equivalent resistance of the parallel connection of $R_c$ and $R_L$, and by $R_{be}$ the resistance base-emitter. Note that the collector voltage is equal to VCC-$i_c R_c$, Its variable component is $v_{out} = -i_c R_{cL}$. The output voltages for the AC component is

$$v_{out} = -i_c R_{cL} = \beta R_{cL} i_b = -\beta R_{cL} V_{in}/R_{be}.$$

So, the gain of the amplifier is

$$K = -\beta \frac{R_{cL}}{R_{be}} \tag{2.1}$$

For example, if $\beta = 50,\ R_{be} = 500\,\text{Ohm},\ R_{cL} = 1\,\text{kOhm}$, then the voltage gain of our amplifier is equal to 100.

Finally, let see that the input resistance (viewed from the input terminal) is equal to $R_{be}$. The output resistance of the amplifier without load, is equal to $R_c$.

Figure 2.2 shows a graphic interpretation of the amplifier design. We can see the curves that represent constant base current on the $V_c - i_c$ plane. Suppose that $R_L = \infty$ (no load resistance). On this plane we can draw the line with slope $1/R_c = di_c/dV_c$. Of course, if $i_c = 0$ then the line reaches the point VCC. On the other hand, if the base current is big, the transistor approaches a short circuit (saturates). Then, the line approaches (approximately) the point ($V_c = 0, i_c = i_c max$). The output voltage oscillates around the operation point as shown in the figure. If $R_L > 0$, then the slope of the load line changes, as shown by the dotted line. However, the operation point does not change because it depends on the DC current component.

An advantage of this amplifier is the high gain K. However, it has some disadvantages:

* The input impedance is low. This will result in considerable current pulled out of the signal source.

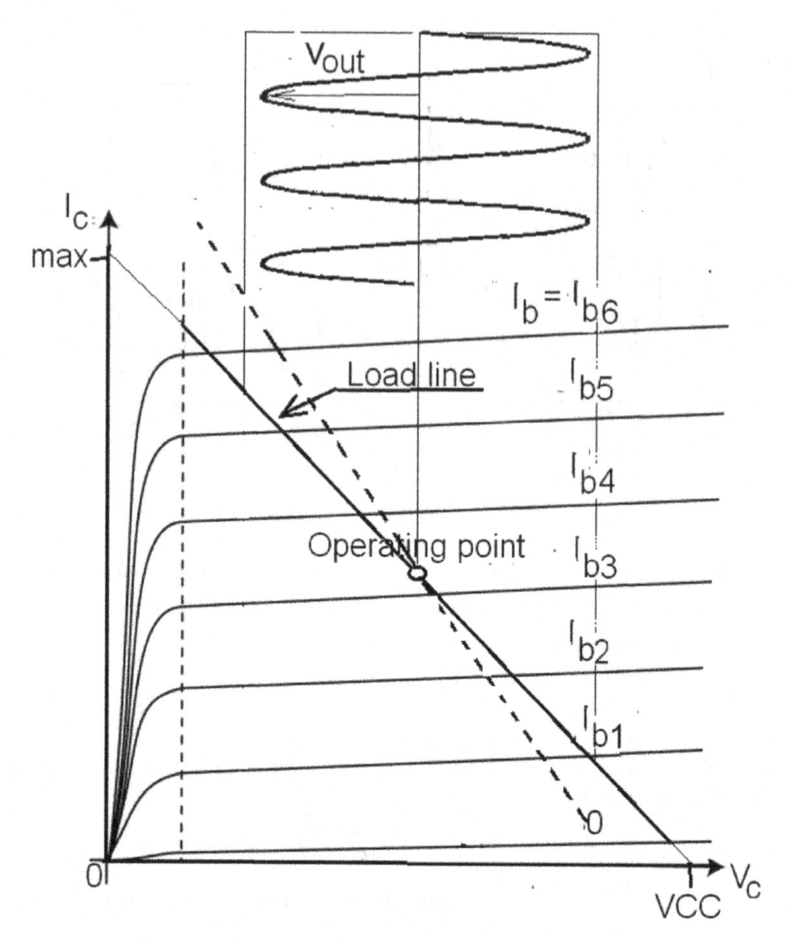


**Fig. 2.2** Transistor characteristics and the load line

* The circuit may reveal certain thermal instability. If the temperature grows, the transistor parameters may change, and $\beta$ may grow. This may cause the change in the operation point.

In the following section we discuss an enhanced version of the voltage amplifier.

## 2.2 Enhanced Amplifier

In Fig. 2.3 we can see the enhanced version of the amplifier of Fig. 2.1. Note that there is an additional resistance between the emitter and the ground, and an optional capacitance has been added in parallel with this resistor. This circuit has somewhat self-stabilizing property. If unwanted collector current growth occurs for any reason, then the voltage on $R_e$ grows. This makes the base current decrease, and helps keep the desired operation point.

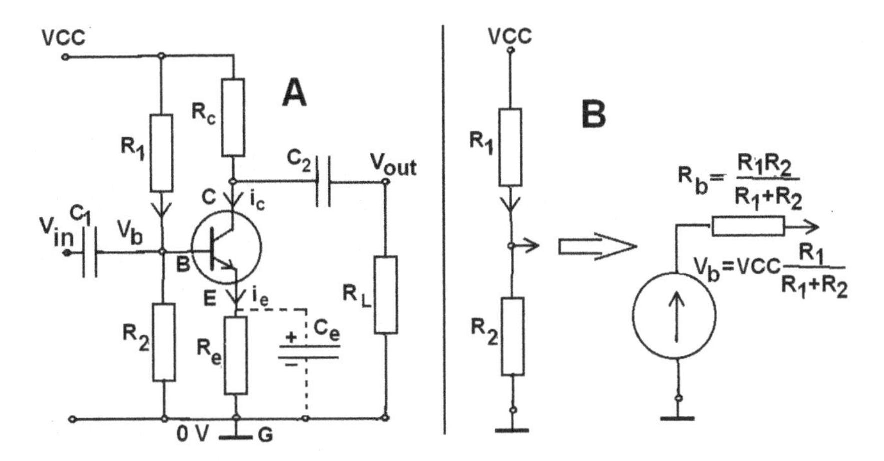


**Fig. 2.3** Enhanced amplifier

The base current is now defined by the voltage divisor $R_1$, $R_2$. This divisor is equivalent to a voltage source, as follows.

$$V_b = VCC\frac{R_1}{R_1 + R_2}, \tag{2.2}$$

with internal resistance

$$R_b = \frac{R_1 R_2}{R_1 + R_2}. \tag{2.3}$$

To calculate $R_c$ we must use voltage VCC reduced by the voltage on $R_e$. So, we have

$$R_c = \frac{VCC/2 - i_{co}R_e}{i_{co}}.$$

This maintains the DC operation point approximately at half of VCC. The initial choice of $R_e$ is arbitrary. After calculating the gain K, we can change $R_e$ and repeat the procedure. In general, $R_e$ should be a small resistance, being a fraction of $R_c$. To keep the operation point, we should have $I_b = i_{co}/\beta$.

The base current is

$$I_b = \frac{i_{co}}{\beta} = \frac{V_b - i_{co}R_e - 0.6}{R_b + R_{be}} \quad (DCcomponent) \tag{2.4}$$

In other words,

$$V_b = VCC\frac{R_1}{R_1 + R_2} = (R_b + R_{be})\frac{i_{co}}{\beta} + i_{co}R_e + 0.6. \tag{2.5}$$

where $R_b = (R_1 R_2)/(R_1 + R_2)$.

**Fig. 2.4** Equivalent AC circuit

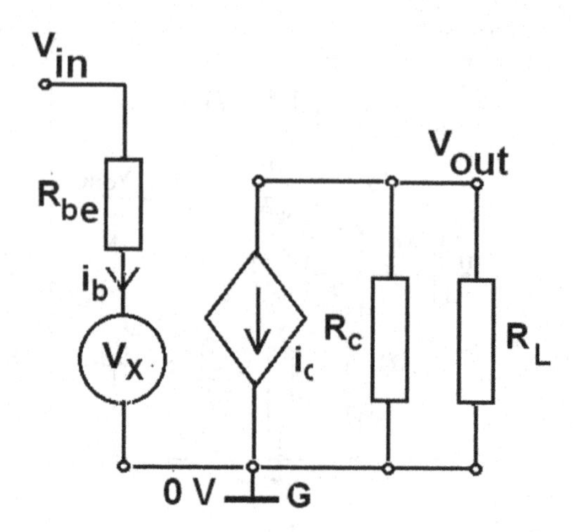


In Eq. (2.5) we have two unknowns: $R_1$ and $R_2$. So, we can choose one of them, and calculate the other (see Eq. (2.2) and (2.3)) . We should select $R_1$ equal or less than (maybe half) of $R_b$ calculated for circuit of Fig. 2.1. Note that $R_1$ and $R_2$ influence the input resistance for the AC component of the input signal.

Now, calculate the gain K for the AC component. First, we assume that the capacitance $C_e$ is not present. Figure 2.4 shows the equivalent circuit for the variable signal component. The voltage $v_x$ is equal to $R_e i_c$.

So, we have

$$\beta i_b = \beta(v_{in} - R_e i_c)/R_{be} = i_c, \tag{2.6}$$

and, after reordering,

$$i_c = \frac{\beta v_{in}}{R_{be} + \beta R_e}. \tag{2.7}$$

Here, we assume big $\beta$ and $i_e \approx i_c$.

Thus,

$$v_{out} = -i_c R_{cL} = -v_{in} \frac{\beta}{R_{be} + \beta R_e} R_{cL}.$$

where $R_{cL}$ is the parallel connection of $R_c$ and $R_L$. This means that

$$K = \frac{v_{out}}{v_{in}} = -\frac{\beta R_{cL}}{R_{be} + \beta R_e}. \tag{2.8}$$

Note that for big $\beta$, the gain is equal to $R_{cL}/R_e$ and does not depend on $\beta$. This means that the gain $K$ is more stable and independent on eventual changes of $\beta$ caused, for example, by fluctuations of the temperature. The addition of $R_e$ also decreases possible non-linear distortions.

Now, consider the circuit with $C_e$ connected. This should be relatively big capacitance, with impedance $|C_e jw| << R_e$. In this case, the equivalent circuit for variable component is equal to that of Fig. 2.1 because the capacitor is a short-circuit for the AC component. Consequently, the gain is $K = -\beta R_{cL}/R_{be}$, like in the amplifies of the previous section.

Observe that both amplifiers discussed above invert the phase of the amplified signal.

The input resistance of the enhanced amplifier can be calculated as follows. From (2.6) we obtain (without capacitor $C_e$)

$$i_b = (v_{in} - R_e i_c)/R_{be} = (v_{in} - R_e \beta i_b)/R_{be}, \tag{2.9}$$

what means that

$$R_{in} = R_{be} + \beta R_e. \tag{2.10}$$

$R_{in}$ is the input resistance viewed from the input terminals $v_{in}$ and ground. Again, if $R_e = 0$, then the input resistance is $R_{be}$, as for the circuit of Fig. 2.1. However, if $R_e > 0$ and for big $\beta$, the input resistance can be considerably big, which is a good circuit property.

## 2.3 Two-Stage Amplifier

Let us connect two (enhanced) amplifiers like those of Fig. 2.3. Instead of $R_L$ of the first amplifier we connect the input of the second stage (Fig. 2.5). The load of the first stage is equal to the input resistance of the second. Suppose that we use the capacitance $C_e$ in the second stage. The total gain of such amplifier may be considerably great, up to several thousand times. This may result in transistor saturation, signal distortion and circuit instability.

Now, connect the output terminal of the second stage with the emitter of the first transistor $T_1$ (close switch S). We assume $R_f >> R_{c2}$.

The current on $R_e$ is now equal to $i_e + i_f$, where $i_f = v_{out}/(R_f + R_e) \approx v_{out}/R_f$. Note that the voltage B-E of $T_1$ decreases by

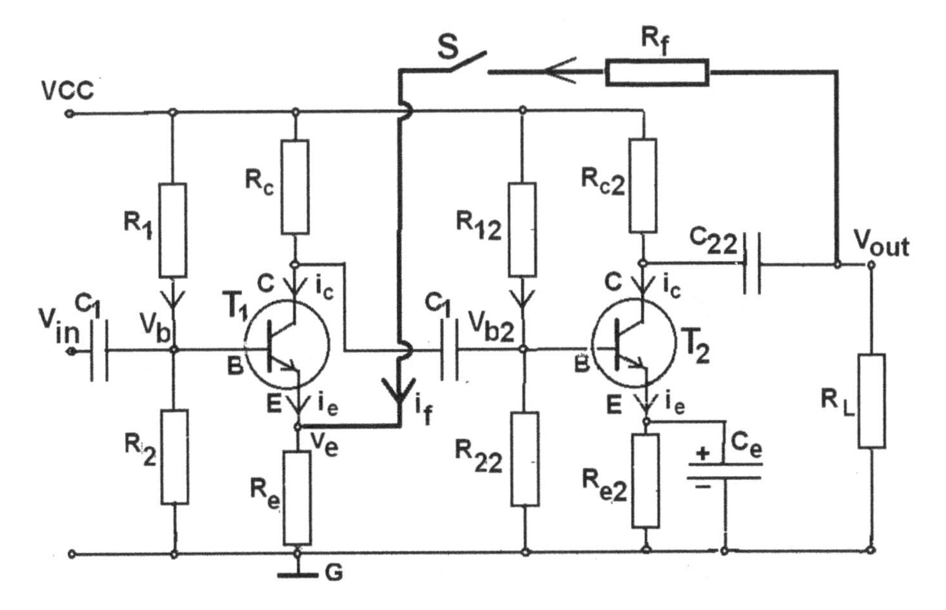


Fig. 2.5 Two stage amplifier

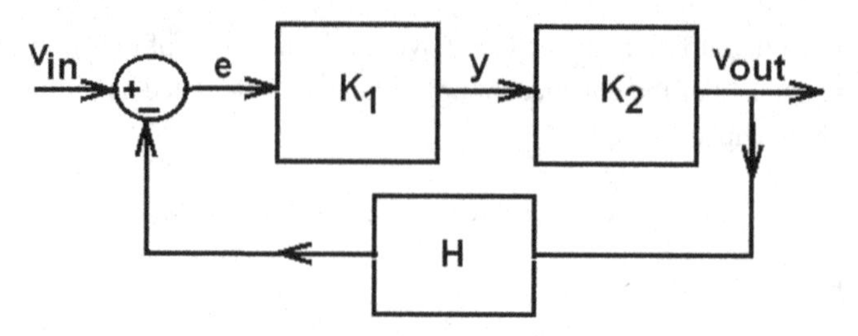


**Fig. 2.6** Block scheme

$$i_f R_e = v_{out} R_e / R_f. \tag{2.11}$$

To assess the total gain $K_t$, look at the generalized block scheme of the circuit in Fig. 2.6. where $e = v_in - Hv_out$. From (2.11) we have $H = R_e/R_f$. Thus, we obtain

$$\begin{cases} v_{out} = eK_1K_2 = \left(v_{in} - v_{out}\frac{R_e}{R_f}\right) K_1K_2 \\ K_{total} = \frac{v_{out}}{v_{in}} = \frac{K_1K_2}{1 + K_1K_2R_e/R_f)} \end{cases} \tag{2.12}$$

Recall that the product $K_1K_2$ is big, so we can neglect the "1" in the denominator. This gives

$$K_{total} \approx \frac{R_f}{R_e} \tag{2.13}$$

On the other hand, if $R_f \to \infty$, then the total gain approaches $K_1K_2$.

With finite $R_f$ and big $\beta$ of the transistors, the gain (2.13) practically does not depend on $\beta$. This amplifier is stable and has very low signal distortion.

## 2.4 The Follower

This is a simple and useful circuit. It does not amplify voltage. The output voltage is approximately equal to the input voltage. The great advantage of the follower is its high input resistance and low output resistance.

Figure 2.7 shows the follower scheme (a) and the equivalent circuit (b) for the signal variable component. The voltage divider $R_1$, $R_2$ should be designed to provide voltage $V_b = VCC/2 + 0.6$. Below, $R_{be}$ is the resistance of junction base-emitter.

From Fig. 2.7 we have:

$$i_b = \frac{v_{in} - v_{out}}{R_{be}}$$

$$v_{out} = R_e(i_c + i_b) = R_e i_b(\beta + 1)$$

$$v_{out} = R_e(i_c + i_b) = R_e\frac{v_{in} - v_{out}}{R_{be}}(\beta + 1)$$

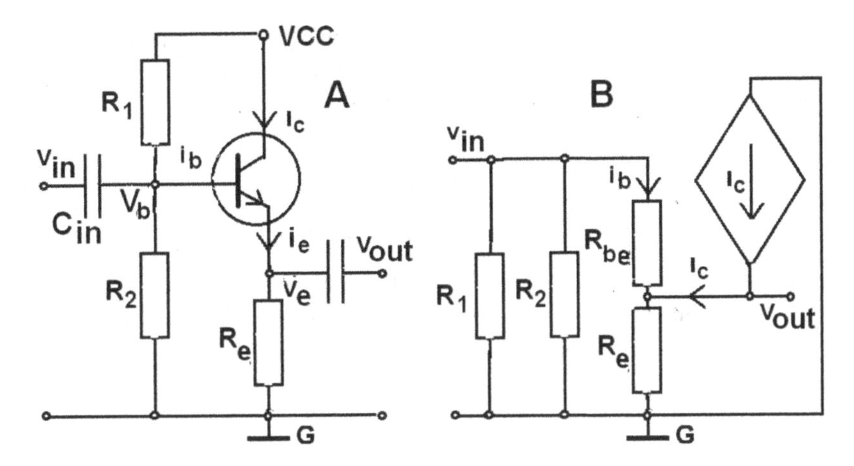


**Fig. 2.7** The follower

After reordering, the gain is

$$K_{\frac{v_{out}}{v_{in}}} = \frac{R_e}{R_{be}} \frac{\beta + 1}{1 + (\beta + 1)R_e/R_{be}}$$

$$K = \frac{\beta + 1}{R_{be}/R_e + \beta + 1} \tag{2.14}$$

Observe that normally, $R_e >> R_{be}$ and $\beta >> 1$. Thus, the approximate gain is equal to one.

Take a look on input and output resistance.

First, ignoring $R_1$ and $R_2$, the input resistance is

$$R_{in0} = \frac{v_{in}}{i_b} = \frac{v_{in}}{\frac{v_{in} - v_{out}}{R_{be}}} = R_{be} \frac{v_{in}}{v_{in} - v_{out}}$$

So,

$$R_{in0} = R_{be} \frac{1}{1 - K}$$

The total input resistance is the parallel connection

$$R_{in} = R_{in0} || R_1 || R_2$$

For big $\beta$, the gain K approaches one, so the resistance $R_{in0}$ is very big. Using big values of $R_1$ and $R_2$, we get a high input resistance. However, remember that the divider $R_1$, $R_2$ must provide the proper constant component of base current. If the signal source provides the constant component approximately equal to VCC/2, then we need no divider at all, and no capacitor $C_{in}$. In this case, the input resistance may be very high.

Now, calculate the output resistance. If no load is connected to the output, then the output voltage is $v_{out} = K v_{in}$. Making short-circuit between the output terminal and the ground, we obtain the output current $i_{short} = i_c = \beta v_{in}/R_{be}$

Finally, he output resistance is

$$R_{out} = \frac{v_{out}}{i_{short}} = \frac{K R_{be}}{\beta} \approx \frac{R_{be}}{\beta} \tag{2.15}$$

For example, if $\beta = 50$ and $R_be = 500$ [Ohm], then the output resistance will be equal to 10 [Ohm]. This is a good property of the follower. If we use the follower as the output stage of a signal processing device, then the signal we send out is resistant to possible external disturbances.

## 2.5 Differential Amplifier

Figure 2.8a shows the scheme of the basic differential amplifier. It has two input terminals where the voltage signals $u_1$ and $u_2$ are applied, and two output terminals $v_1$ and $v_2$. There are two modes of operation: the *common mode*, where the two input signals are equal to each other, and the *differential mode* where different inputs are applied, and the result is the amplification of the difference $u_1 - u_2$.

The common emitter resistance need not be small, as in the enhanced amplifier of Sect. 2.2. It is reasonable to choose $R_e = R/2$. Below, we discuss the gain for variable component of the signals.

First, consider the common mode operation. In this case, $u_1 = u_2$. The transistors are identical to each other, with identical collector resistances $R$. Consequently, we can replace the circuit with only one transistor.

The gain of the circuit of Fig. 2.8b has been calculated before, see Eq. (2.8). In our case, it is as follows (Eq. (2.16)),

$$K_{com} = \frac{v}{u} = -\frac{\beta R}{R_{be} + \beta R_e} \tag{2.16}$$

where $R_{be}$ is the resistance base-emitter.

We can see that the gain $K_{com}$ is rather small, and for big values of parameter $\beta$ it approaches $R/(2R_e)$.

Now, apply to the inputs different voltages. Denote $u = u_1 - u_2$. Assume $u_1 = u_{com} + d_u$, $u_2 = u_{com} - d_u$. If $d_u = 0$, we have the common mode operation, discussed earlier. So, now consider the "pure differential" input, with $u_{com} = 0$, $d_u > 0$.

Observe that is this case the current of $R_e$ remains constant (variable component zero) because the changes of $i_{e1}$ and $i_{e2}$ are equal to each other, with opposite sign. So, in the equivalent circuit for variable component the collector voltage is constant, with variable component zero, and the resistance $R_e$ disappears from the equivalent circuit. The two transistors work separately, as a simple amplifiers, with opposite input signals $d_u$ and $-d_u$.

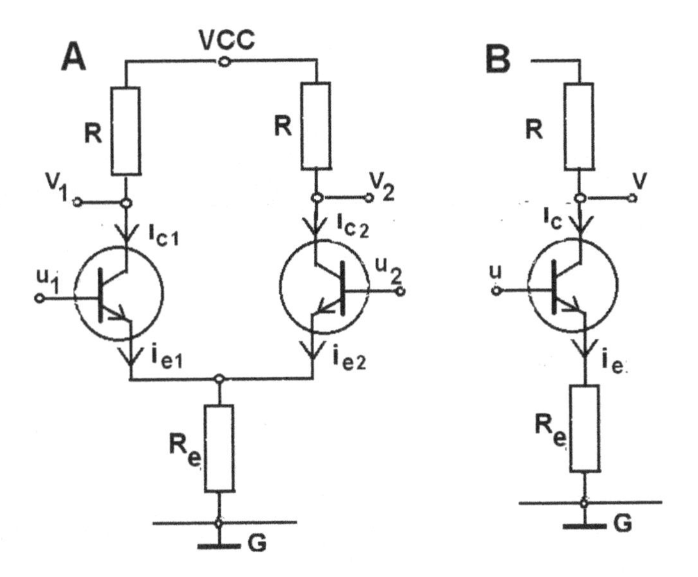


**Fig. 2.8** Differential amplifier

The gain K of such amplifier has been calculated earlier, see Eq. (2.1) This means that $v_1 = -d_u K$ and $v_2 = d_u K$. Thus, we obtain the gain $K_{dif}$ for the difference of the signals as follows (Eq. (2.18)).

$$K_{dif} = \frac{v_2 - v_1}{u_2 - u_1} = -2\beta \frac{R}{R_{be}} \tag{2.17}$$

The *common mode rejection ratio* (CMRR) is the ability of the device to reject common-mode signals. The definition is as follows.

$$CMRR = 20\,log_1 0 \left( \frac{K_{dif}}{K_{com}} \right) \; [dB]$$

For example, if $\beta = 60$, $R = 1\,\text{k}\Omega$, $R_e = 500\,\Omega$ and $R_be = 300\,\Omega$, we obtain $K_{dif} = 600$, $K_{com} = 0.99$, CMRR $= 118.7$ dB.

Now, suppose that $u_{com} > 0$, $d_u > 0$, and that we are working in the linear region of operation of the transistors. So, the resulting voltages can be obtained by the superposition of the voltages of common and differential mode. Finally, we have

$$v_2 - v_1 = K_{dif}(u_2 - u_1), \quad (v_2 + v_1)/2 = v_{com} = K_{com}(u_1 + u_2)/2 \tag{2.18}$$

Note that the two output signals $v_1$ and $v_2$ have opposite phase. So, we can get, as the output, the difference, as well as inverted or not inverted signal.

An amplifier with high CMRR coefficient is useful in many applications. For example, if we send a weak, low voltage signal from a microphone through a large cable, the signal is

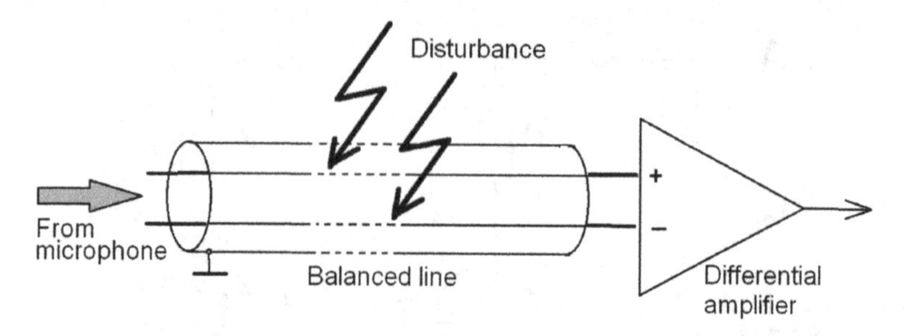


**Fig. 2.9** Balanced signal

subject to disturbances produced by nearby devices. If it were a simple cable with one signal wire, then the disturbance would affect the signal and enter the amplifier. Now, suppose that we use the balanced line, as shown in Fig. 2.9, sending the signal as a voltage difference. It is very probable that the disturbance influences both wires in a similar way. This has little effect on the difference (the useful signal) that is amplified by the differential amplifier.

### 2.5.1 Enhanced Differential Amplifier

Figure 2.10 shows a possible modification of the basic differential amplifier of Fig. 2.8. Instead of the resistance $R_e$ we use a transistor $T_3$. The diodes $D_1$ and $D_2$ are used as a source of (approximately) constant voltage equal to 1.2 V. This stabilizes the voltage $v_1$ of the base of $T_3$. This way, $T_3$ works as a constant source of current. If so, the connection $T_3 - T_o$ can be treated as a huge resistance $R_e$, practically infinite. Note that $R_e$ appears in the denominator of the formula (2.16) for the gain in common mode. This means that the amplifier practically does not amplify the common mode component, and has a huge CMRR coefficient (Fig. 2.10).

In the amplifier of Fig. 2.10, the voltage $v_e$ of the emitters of $T_1$ and $T_2$ is defined by the constant component of the input voltages $u_1$ and $u_2$.

There are many other versions of the differential amplifier. For example, if we use the field effect transistors (FET) instead of the conventional BJT, we obtain an amplifier with very high input impedance.

### 2.5.2 Dynamic Microphone Amplifier

Figure 2.11 shows a scheme of a dynamic microphone. The incoming sound waves move the diaphragm that is connected with a moving coil. The coil vibrates and its wires intersect

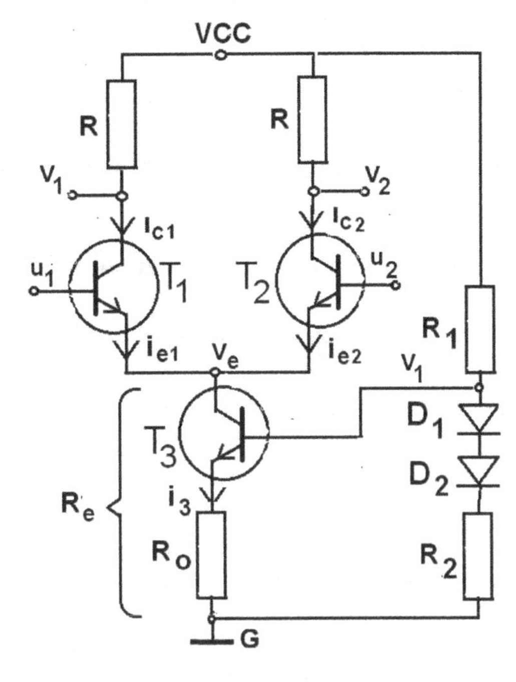


**Fig. 2.10** Enhanced differential amplifier

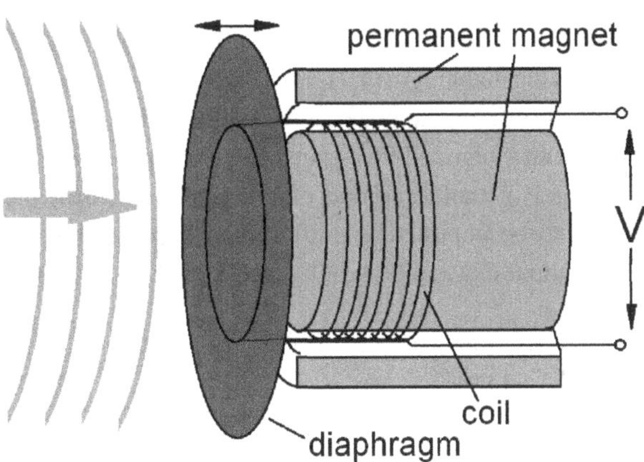


**Fig. 2.11** Dynamic microphone

a strong magnetic field produced by the permanent magnet. This induces the electomotive voltage in the coil according to the sound waves.

A typical dynamic microphone has a coil with impedance of 150–800 Ω and produces several mV of output voltage. A *pre-amplifier* is needed to send and process the microphone signal in other devices. Normally, the microphone signal is send through a balanced microphone cable, see Fig. 2.9. This means that the amplifier should have a differential input. The application of op-amp is a possible solution, but it is somewhat inconvenient. The op-amp needs positive and negative supply voltage, and it cannot be used without negative feedback.

If we add the feedback to determine the gain, the input to the one-op-amp circuit is no longer symmetric. The most appropriate is a one- or two-stage differential amplifier. However, the most practical application is possible using an integrated circuits like LM386. This IC needs few external elements, and works with one, positive supply source. It can be configured to have gain of 20x, up to 200x (26–46 dB).

### 2.5.3 Condenser Microphone Amplifier

The *condenser microphone* is a capacitor with one plate being a thin conductive diaphragm D the moves according to the pressure of the acoustic wave. As the distance between the plates changes, the capacity of the microphone changes. If there is a non-zero charge on the capacitor, then the voltage must change. This way we obtain an electric signal that corresponds to the received acoustic wave (see Fig. 2.12).

A simple circuit with condenser microphone in shown in Fig. 2.13. The resistance provides the charge to the microphone capacitance. The voltage signal from the microphone is sent through the capacitor C to the output. This signal needs to be amplified.

The circuit of Fig. 2.13 is simple, but not very practical. The microphone capacitance is very small, so the charging resistance must be big (several megaohms). This makes the output impedance of the circuit very big, and the output signal becomes highly sensitive to disturbances. A good solution is the *electret microphone*. This device consists of a capacitor with moving diaphragm that is pre-charged using a permanently charged material. It contains a static electric dipole material maintaining the charge practically forever. The microphone comes with integrated FET transistor, as shown in Fig. 2.14.

The gray rectangle contain elements that are inside the electret microphone capsule. The signal may be amplified by a differential amplifier. If so, we can get the differential amplified signal output that can be sent by the balanced microphone cable.

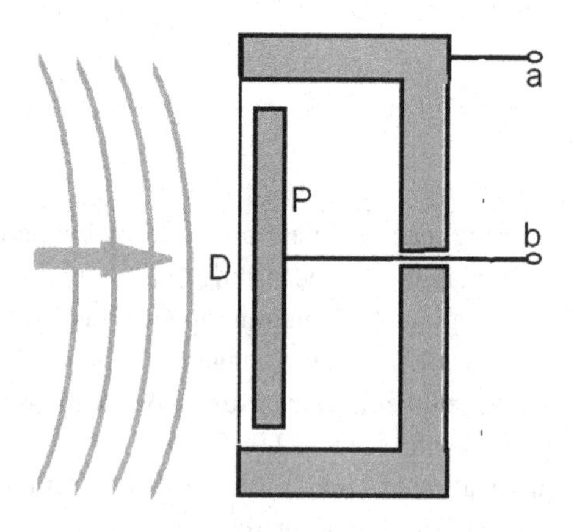


**Fig. 2.12** Condenser microphone

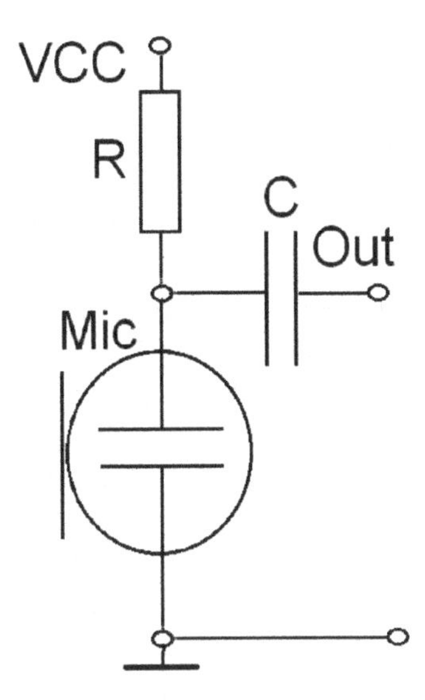


**Fig. 2.13** Simple condenser microphone circuit

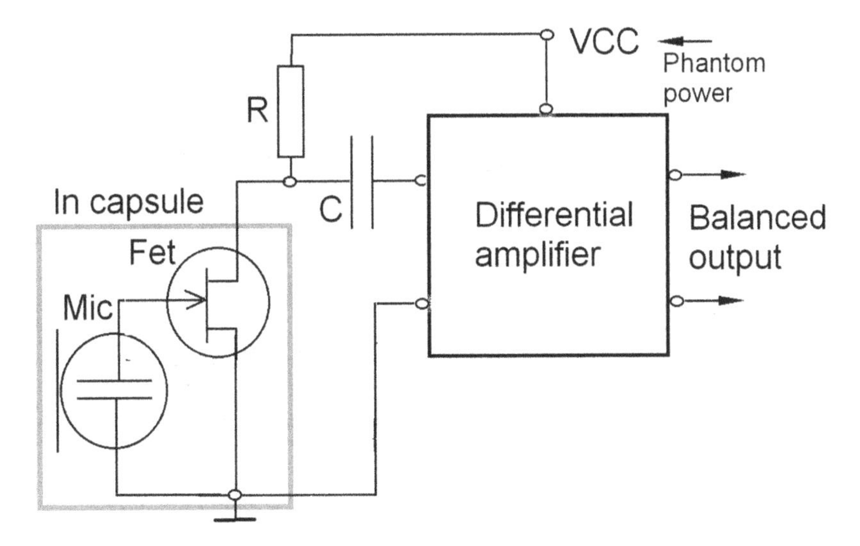


**Fig. 2.14** Electret microphone

The circuit of Fig. 2.14 needs to be supplied by a VCC voltage source. This is done by the *phantom power* voltage that is sent through the same wires of the balanced signal. This voltage is normally between 9 and 50 DC, and is connected to the signal wires by two resistors of 6.8 kΩ. We do not consider here the bad practice of sending the microphone signal through a one-wire cable.

## 2.6 Modulation and "Tremolo"

### 2.6.1 A Simple Modulator

A simple modulator is shown in Fig. 2.15. Suppose that we want to modulate an acoustic signal with other, low frequency sinusoidal voltage. The transistor of Fig. 2.15 works at the (lower) limit of the active linear region. This makes the $\beta$ parameter variable, depending on the position of the operation point. Now, a low frequency (4–10 Hz) is added to the signal. As the result, the gain of the amplifier fluctuates with low frequency, and the amplitude of output signal is changing. This produces a "tremolo" effect in audio applications. The capacitance $C_2$ must have great value, to permit the low frequency to enter at the transistor base.

The cutoff frequency for the output circuit (high-pass filter) $R_3C_3$ must be greater that the modulating (low) frequency $f_m$. So, we must have $R_3C_3 > f_m$. For example, if the modulating frequency is equal to 5 Hz, we must have $1/(2\pi R_3C_3) > 5$. The cutoff frequency for $C_2R_u$ should be greater than 5 Hz, and lower that 20 Hz (the low limit of acoustic signals).

The above circuit may work satisfactory, but it has certain disadvantages. The useful signal may be distorted, and the modulating low-frequency signal may pass to the output terminal. Better results can be obtained using the differential amplifier.

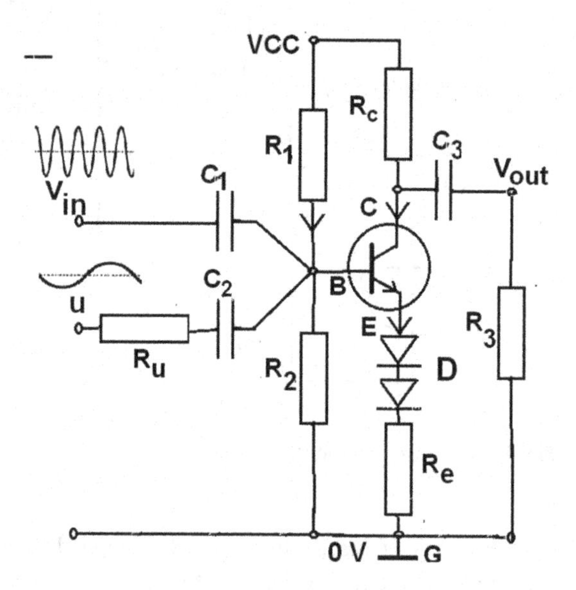


**Fig. 2.15** A simple modulator

### 2.6.2 Modulator with Differential Amplifier

Figure 2.16 shows a modulator with differential amplifier. Here, the modulating, low frequency signal is applied to the base of transistor $T_3$. This transistor defines the operation point for $T_1$ and $T_2$. Again, the transistors should work at the lower limit of the linear active region, where $\beta$ decreases with decreasing DC emitter current. This makes the gain in differential mode change, and the output signal is modulated by the voltage $v_m$. In this modulator, the output must be taken as the difference $v_1 - v_2$. Modulated signal will look like that of Fig. 2.17.

### 2.6.3 AM and FM

The history of radio communication began in 1895, when Gugliemo Marconi invented what he called "the wireless telegraph". First, the Morse code was transmitted. In 1900, first audio signals of speech have been transmitted. Up to the middle of XX century, the radio communications have been based on amplitude modulation (AM) of the carrier high frequency, like the signal of Fig. 2.17.

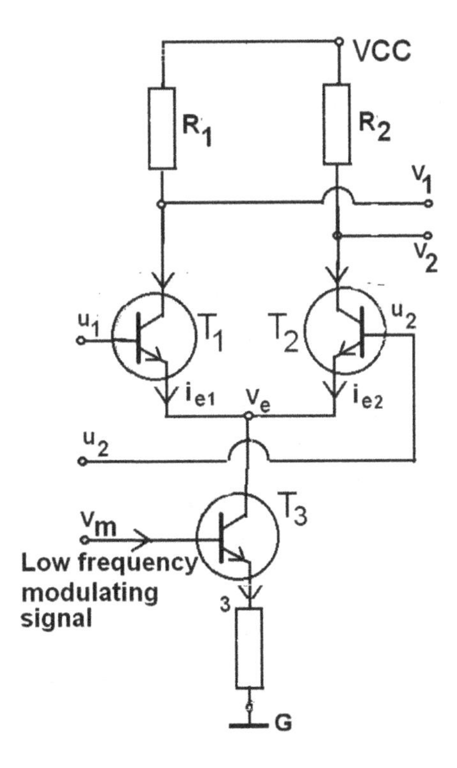


**Fig. 2.16** Modulator with differential amplifier

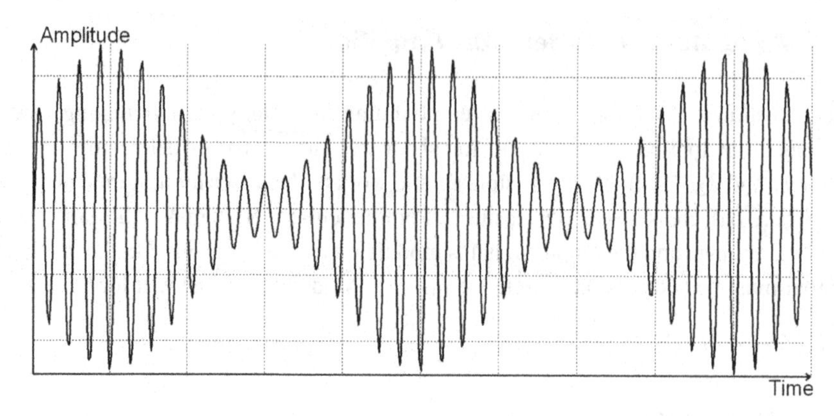


**Fig. 2.17** Modulated signal

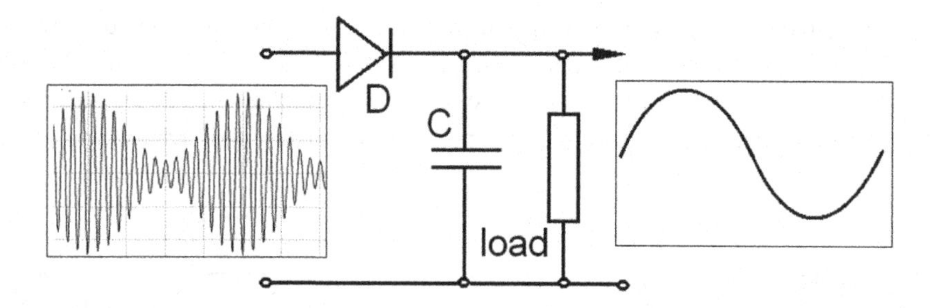


**Fig. 2.18** AM detector

In the AM transmission, the audio signal with acoustic frequency is mixed with the carrier frequency (hundreds of kHz or several mHz) that produce AM-modulated radio signal. At the receiver, this signal is rectified to get the audio frequency, and the carrier frequency is filtered out. The obtained signal may be amplified and passed to the speaker. A simple diode detector that does the job is shown in Fig. 2.18.

Note that in the AM, as well in FM, the carried frequency does not carry any information. The broadcast information appears when the frequency is being modulated. This, however, produces other "lateral" frequencies that must be sent and detected by the receiver. Take a look at the AM modulated signal of Eq. (2.19), where $\omega_L$ is the low-frequency modulating signal, and $\omega_H$ is the high, carrier frequency.

$$f(t) = sin(\omega_H t)\, sin(\omega_L t) \tag{2.19}$$

Using the formula for the product of two sinusoidal functions, we have

$$f(t) = \frac{cos(\omega_H - \omega_L) - cos(\omega_H + \omega_L)}{2} \tag{2.20}$$

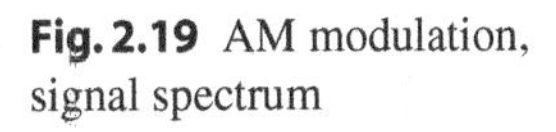

**Fig. 2.19** AM modulation, signal spectrum

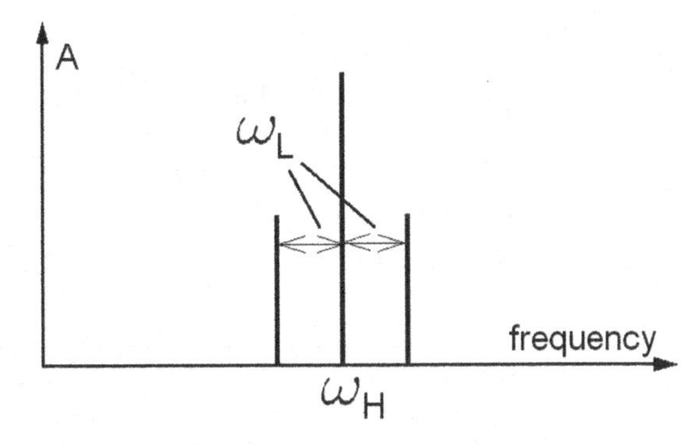


**Fig. 2.20** LC circuit with varicap

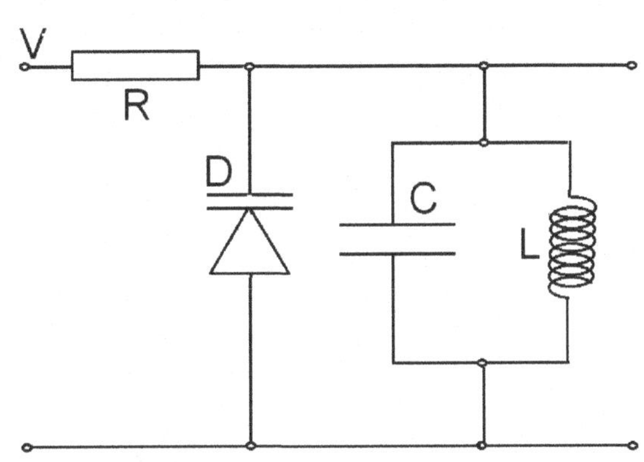


This means that in the modulated signal two different frequencies appear: $\omega_H - \omega_L$ and $\omega_H + \omega_L$, see Fig. 2.19. So, instead of one carrier frequency certain frequency range is being sent and must be received. This means that the filter(s) that select the received radio frequency cannot be too selective, to receive not only the carrier frequency but also the lateral frequency range.

One of the great disadvantages of AM transmission is the sensitivity to external disturbances, like static discharges or other electric devices that work in the environment. In AM, such disturbance affects directly the signal amplitude and enter the receiver. To avoid the problems with AM interference, the frequency modulation (FM) has been applied. Instead of modulate the amplitude, the FM high frequency carrier wave has the frequency modulated by the audio signal. To modulate the frequency of an LC oscillator, we can use the varicap (variable capacitance) diode (see Sect. 1.10.1).

A simple application of the varicap diode is shown in Fig. 2.20, The varicap works in parallel with the capacitance C, and the two capacitors determine the frequency. The controlling low frequency signal V(t) is applied to the varicap, with the necessary DC positive bias. This circuit can be used in any LC-based high frequency oscillator, to generate the FM signal.

Figure 2.21 depicts an example of frequency modulated signal.

The detector of Fig. 2.18 cannot extract the modulating (low) frequency from the FM signals. Circuits that are used for FM detection are referred to as *phase discriminators*. Figure 2.22 shows an example of such circuit.

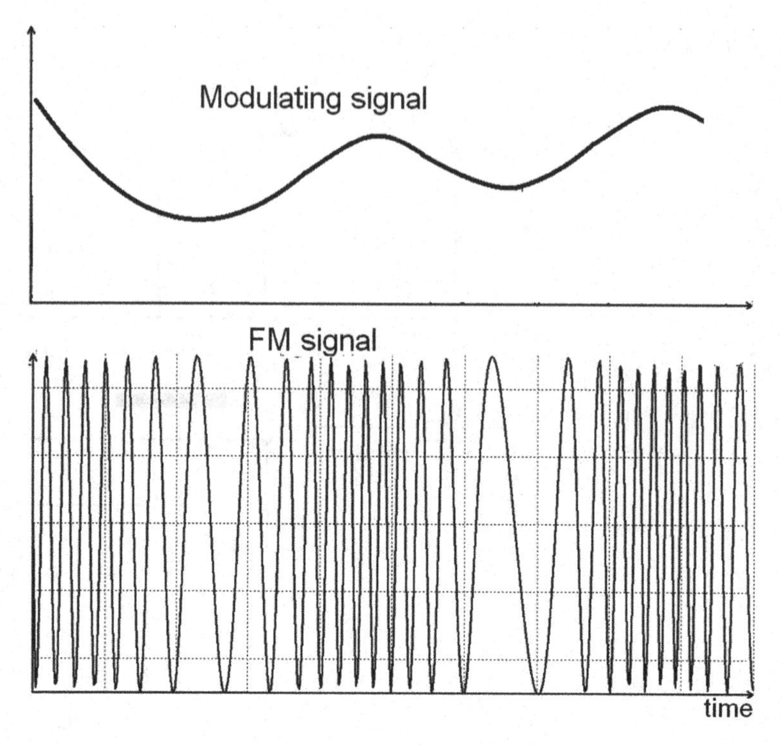


**Fig. 2.21** FM signal

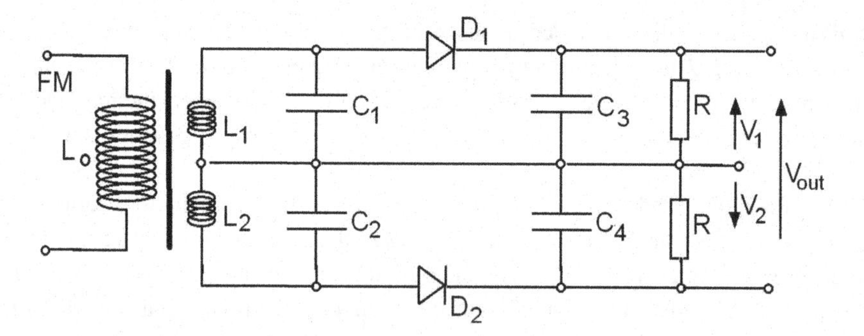


**Fig. 2.22** Phase discriminator

There are three coupled inductances $L_o$, $L_1$ and $L_2$, The FM signal is applied to $L_0$ and then it is induced on the other inductors. As we can see, there are two rectifiers with diodes $D_1$ and $D_2$, with opposite polarity. The resulting voltage $V_{out}$ is the difference $V_{out} = V_1 - V_2$. The two LC circuits are tuned to slightly different frequencies, $f_1 < f_2$. So, the circuit $L_1, C_1, D_1$ and $L_2, C_2, D_2$ produce rectified low frequency voltage with displaced maximal values, as shown in Fig. 2.23. The superposition of these voltages, $V_{out}$ is shown at the lower

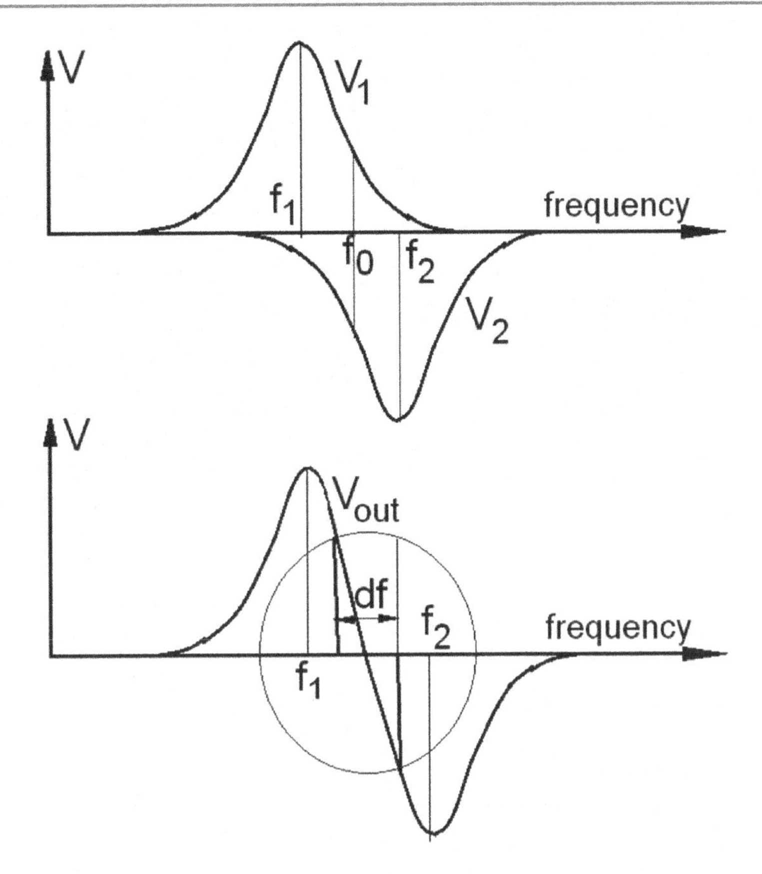


**Fig. 2.23** Phase discriminator frequencies and voltages

part of the figure. There is an interval of the FM frequency, marked as "df", where the output voltage depends in almost linear way on the frequency. This region is used to convert the frequency fluctuations into the output voltage. Before applying the FM signal to $L_0$, it is amplified and passes through a limiter that makes the amplitude constant, eliminating the eventual AM disturbances.

## 2.7 Low Noise Amplifier

The unwanted noise is always present in signal processing devices. There are many possible sources of noise. These may be fluctuations of the voltage on nearby devices, high frequency radio signals captured by some part of our amplifier, a "static" potential accumulated on things, and the noise generated by the transistors and other parts of our circuit. The influence of external noise can be decreased by good insulation and screened cables. However, the noise generated on semiconductors, electrolytic capacitances and other devices of the circuit is difficult to avoid.

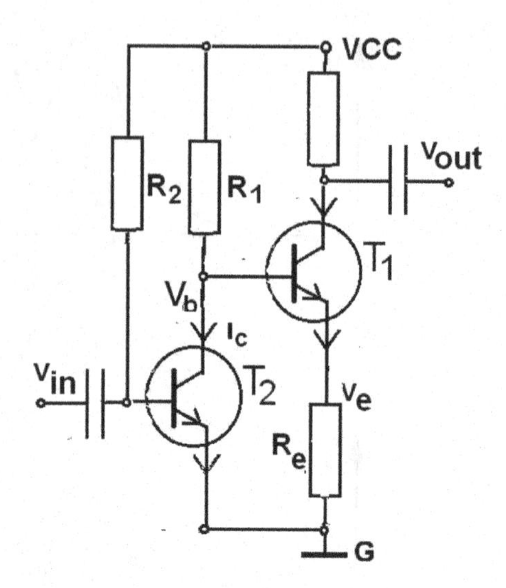


**Fig. 2.24** A low-noise amplifier

The most frequent source of noise is caused by the fluctuation of current on NP or PN junction of semiconductors. In a transistor, such noise is mostly produced on the collector-base junction because normally this is where we have a big potential difference. What we need is to reduce the noise/signal ratio. While amplifying signals of very small amplitude, provided by sources of big internal resistance, this ratio may be unacceptable.

Figure 2.24 shows a possible solution.

In this amplifier, there is an additional transistor $T_2$, connected to the base of the amplifying transistor $T_1$. It can be seen that the collector voltage of this transistor is equal to the base voltage of $T_1$, so, it is only a fraction of VCC. The voltage is approximately 0.6 V plus the voltage drop on $R_e$, that is normally a small resistance. This means that $T_2$ must be a transistor for low voltage and current. Moreover, this transistor may work at the lower limit of its operation region. The eventual non-linearity of $T_2$ will not cause big distortion because it is supposed that the input signal AC voltage is very low. The low voltage on $T_2$ means that there is little noise produced on the collector-base junction. Anyway, $T_2$ amplifies the signal, and the variable component of its collector current $i_c$ is the controlling signal for $T_1$. It should be $R_2 >> R_1$, and both resistances should be great, the $R_2$ of one or more $M\Omega$.The low voltage on $T_2$ makes the signal-to-noise ratio small.

## 2.8 Operational Amplifier

By *operational amplifier* (op-amp), we mean an amplifier with differential input stage, huge gain (more than several thousand times), which amplifies the variable, as well the constant component of the input signals. Figure 2.25 shows the symbol of an operational amplifier. Note that the op-amp operates with positive and negative power supply. So, it can amplify positiva, as well negative signals. If the signal is equal to zero, the output voltage should also be zero. Normally, op-amps work with VCC = ±15 V.

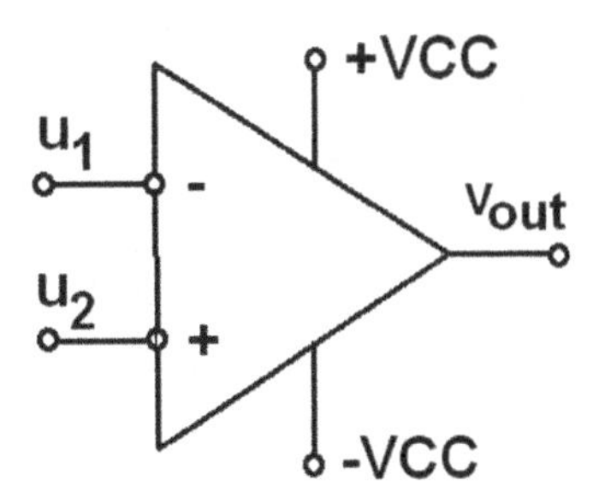


**Fig. 2.25** The symbol of the operational amplifier

In the amplifier of Fig. 2.26 we can see a basic op-amp scheme.

There are two power supply sources: +VCC and -VCC. The ground terminal G has voltage equal to zero.

The amplifier has two differential stages, the first formed by $T_1$, $T_2$ and $T_3$, and the second with $T_4$ and $T_5$. The output stage, with transistors $T_6$, $T_7$, $T_8$ and $T_9$ is a "push-pull" power amplifier that will be discussed further on.

The first stage is similar to the enhanced differential amplifier discussed in Sect. 2.6.1. The second stage is a simple differential amplifier, where the input signals are taken from the collectors of $T_1$ and $T_2$ by the direct connection. The signal from the collector of $T_6$ controls the output, power stage. Observe that the first differential amplifier works between + and - VCC, and the second is placed between the ground (V $= 0$) and +VCC. This permits to connect directly the stages, without any need of separating capacitances. Observe also that the transistor $T_4$ has no collector resistor and is connected directly to +VCC. This is done because there is no need to use the signal from collector of $_4$, and the eventual resistor placed there would have no influence on the current of $T_4$ and its operation point (Fig. 2.26).

The connection $R_4 - D_2$ provides the voltage $v_{b7}$ that is stabilized by the diode. This converts the transistor $T_7$ in a constant current source, needed in the push-pull stage. The operation of the push-pull will be explained in the following section.

The amplifier has two input terminals $u_1$ and $u_2$. Suppose that the voltage of signal $u_1$ grows. The small arrows indicate the resulting voltage changes in the consecutive points of the circuit. Finally, the change of the output signal is negative. This means that the amplifier inverts the phase of signal $u_1$. The other input terminal is non-inverting.

Operational amplifier has a very high voltage gain, and should never be used without a feedback circuit.

Figure 2.27 shows an operational amplifier with negative feedback. The signal from the output terminal is passed to the negative input through the resistance $R_1$. The voltage divisor $R_1 - R_2$ defines the voltage $w$ on the negative input, equal to $u_1 + (v_{out} - u_1)R_2(R_1 + R_2)$. Recall that the op-amp has a very high input resistance, so the voltage $w$ depends only on $R_1$, $R_1$ and $v_{out} - u_1$.

The coefficient K is the gain of the amplifier. We have

$$\begin{cases} w = u_1 + (v_{out} - u_1)\frac{R_2}{R_1+R_2} = v_{out}\frac{R_2}{R_1+R_2} + u_1\frac{R_1}{R_1+R_2} \\ w = \frac{v_{out}}{K} \end{cases} \tag{2.21}$$

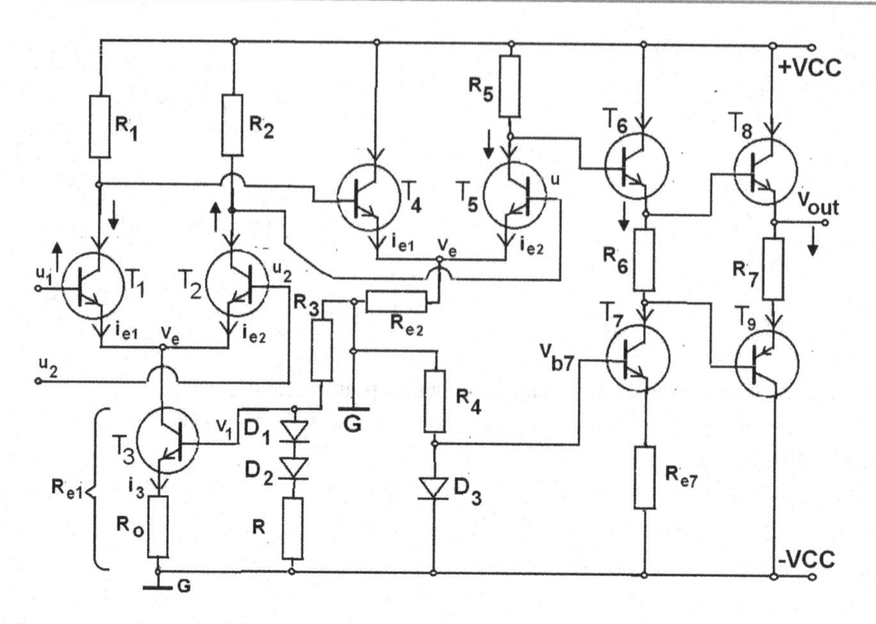


**Fig. 2.26** Operational amplifier, simplified scheme

**Fig. 2.27** Operational amplifier with feedback

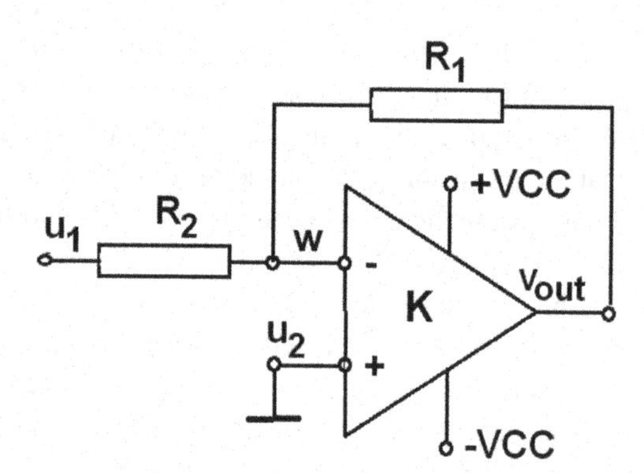


The gain K is practically infinite, so, if $v_{out}$ is finite, we have $lim_{K \to \infty} w = 0$. Thus, from the first equation of (2.21) we obtain the total gain G with feedback as follows.

$$G = \frac{v_{out}}{u_1} = -\frac{R_1}{R_2} \tag{2.22}$$

The op-amp with feedback is a highly lineal and stable device. Note that the gain G does not depend on K, provided K is big. In the next chapter, we will discuss op-amps with feedback formed by complex impedances. This can be used in designing filters and signal processing instruments.

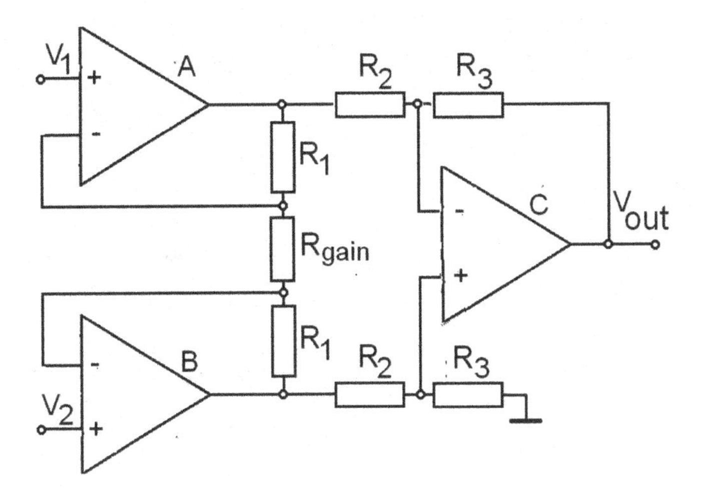


**Fig. 2.28** Instrumentation amplifier

### 2.8.1 Instrumentation Amplifier

If we use an op-amp with feedback like that of Fig. 2.27, we get a highly linear and stable voltage amplifier. However, in instrumentation and signal processing, we frequently need an amplifier with differential input. This can be done, by using the other (positive) input of the opamp, instead connecting it to the ground. This may work, but observe that the inputs will not have the same characteristics. While the positive input will have a very input impedance, the input resistance will be equal to $R_2$.

A good solution is the *instrumentation amplifier* made of three op-amps, as shown in Fig. 2.28.

If $R_{gain} = \infty$, then the amplifiers A and B work as followers. If $R_{gain}$ is finite, it defines the overall gain of the circuit. The inputs are completely symmetric, each of them with very high input impedance. The gain of the whole amplifier is given by Eq. (2.23).

$$K = \frac{V_{out}}{V_2 - V_1} = \left(1 + \frac{2R_1}{R_{gain}}\right)\frac{R_3}{R_2} \tag{2.23}$$

## 2.9 Power Amplifier Stage

Consider a simple voltage amplifier, like that of Fig. 2.3. Now, we want to connect a speaker or other power-consuming devise (Fig. 2.29).

The first idea is to connect it instead of the collector resistance. However, the speaker should not receive the DC component in steady state (silence), and should be powered only by the AC component. So, we connect it instead of the load $R_L$. This may work, but from

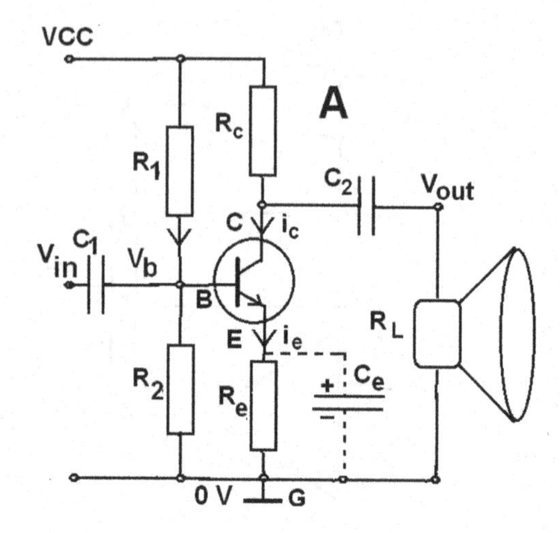


**Fig. 2.29** Connecting speaker

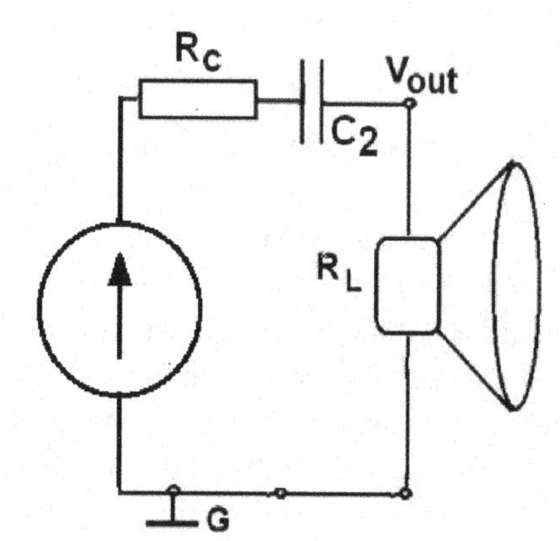


**Fig. 2.30** Equivalent circuit

the point of of efficiency, this is not correct. The DC current of the transistor must be great enough to ensure the desired AC power on the load. This means that we will lose a big part of power at the transistor and the collector resistance (Fig. 2.30).

Another possibility is to connect the speaker at the collector using a transformer (see Fig. 2.31). This may work, but the amplifier will have rather low efficiency. What we design is the power amplifier of class A. Suppose that the (ideal) transformer Tr has the ratio $n_2/n_1 = 1$ and that we amplify a sinusoidal signal of frequency $f$. Observe that the DC impedance of the primary part of the ideal transformer is equal to zero, so the average voltage $v_c$ is equal to $VCC$. Thus, for the maximal signal amplitude, the voltage $v_c$ oscillates between zero and $2VCC$. Denote shortly VCC as V. So, the effective voltage on the load is $v_{eL} = V/\sqrt{2}$.

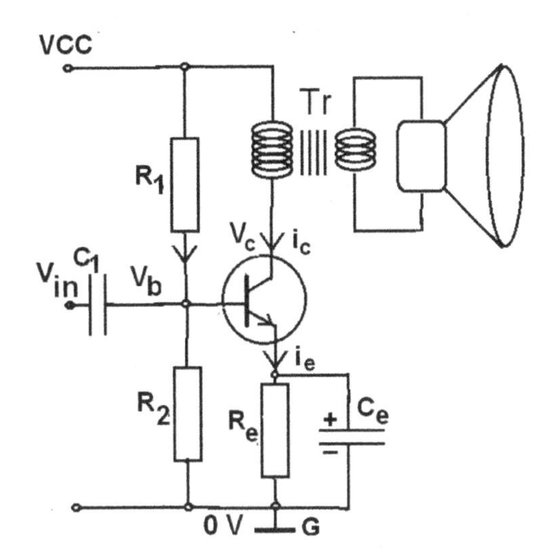


**Fig. 2.31** Using transformer. Amplifier of class A

Consequently, the power on the load is

$$P_L = (v_{eL})^2/R_L = \frac{V^2}{2R_L}. \tag{2.24}$$

The load voltage and current are as follows, respectively: $v_l(t) = Vsin(\omega t)$, $i_L(t) = Vsin(\omega t)/R_L$. As the transformer turns ratio is equal to one, the AC component of collector current $i_{ac}(t) = i_L(t)$, and the current is $i_c(t) = Vsin(\omega t)(sin(\omega t + 1))/R_L$ .

On the other hand, it is easy to see that the supplied power is equal to $V^2/R_L$. This means that the efficiency of the amplifier is equal to $\eta = 50\%$. This (theoretical) efficiency is approached, supposing the input signal that results in maximal oscillation of the collector voltage.

### 2.9.1 Class B Amplifier "Push-Pull"

The idea of the "push-pull" power stage is to use two transistors, providing a positive and negative current on the load, respectively (see Fig. 2.32, part A). In the first half of the signal period, transistor NPN provides the current $i_{c1}$ that pass to the load because the other transistor is closed. In the second part, the current of NPN is equal to zero, and the PNP current is $i_{c2}$. This means that the load current in this period in equal to $-i_{c2}$.

Figure 2.33 shows the load voltage, provided by both transistors. We suppose that the input signal produces maximal oscillation on the load. Denote $VCC = V$. The voltage on the load is $v_L(t) = Vsin(\omega t)/2$ and the current $i_L(t) = Vsin(\omega t)/(2R_L)$. Let us calculate the power on the load. The energy produced during one period of the input signal is as follows.

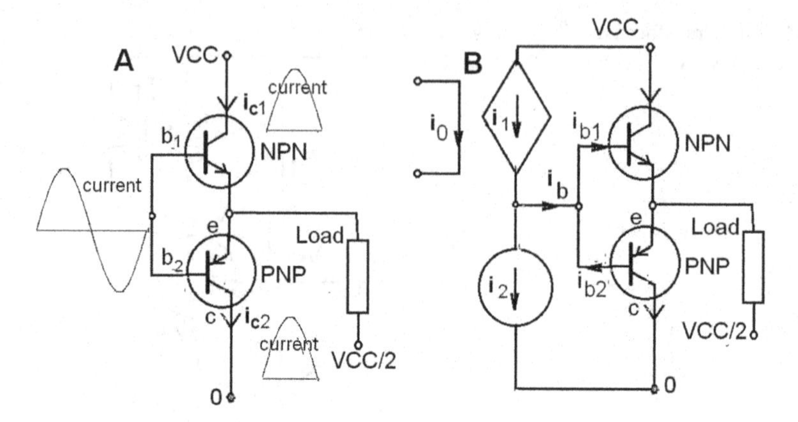


**Fig. 2.32** Amplifier of class B "push-pull"

**Fig. 2.33** Amplifier of class B, load voltage

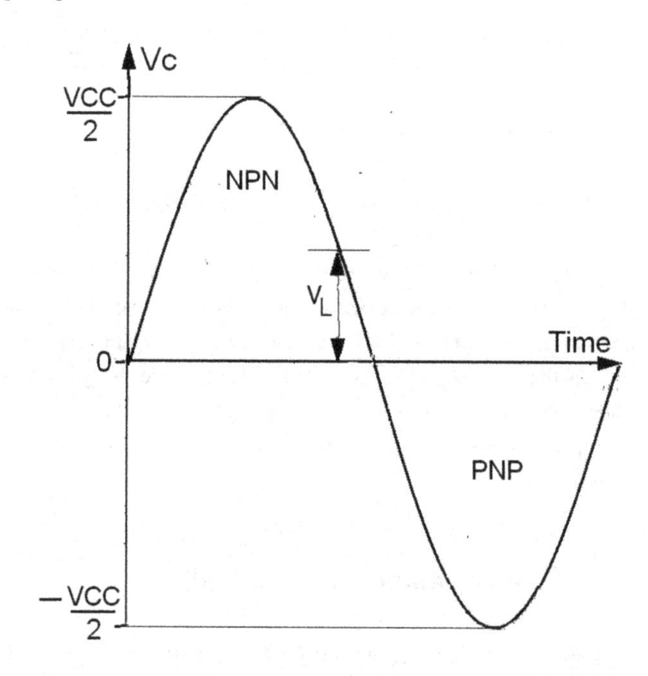


$$E_L = \int_0^T \frac{v_L(t)^2}{R_L} dt = \frac{V^2}{4R_L} \int_0^T sin(\omega t)^2 dt = \frac{V^2 \pi}{4R_l \omega} \tag{2.25}$$

So, the power on the load is

$$P_L = \frac{E_L}{\frac{2\pi}{\omega}} = \frac{V^2}{8R_L} \tag{2.26}$$

The energy and power supplied during one signal period is as follows (note that $i_L(t)$ flows from VCC only during the first half of the period)

$$\begin{cases} E_S = V \int_0^{T/2} i_L(t)dt = V \int_0^{T/2} V sin(\omega t)/(2R_L)dt = \dfrac{V^2}{2R_L} \int_0^{T/2} sin(\omega t)dt \\ E_S = \dfrac{V^2}{\omega R_L} \end{cases} \tag{2.27}$$

and, finally,

$$P_S = \frac{V^2}{2\pi R_L} \tag{2.28}$$

The efficiency of the amplifier is

$$\mu = \frac{P_L}{P_S} \times 100 \approx 78.5\% \tag{2.29}$$

Note that in the idle run with input signal zero, the theoretical power consumption of this amplifier is equal to zero because both transistors are closed.

Figure 2.32, part B depicts a possible control circuit for the push-pull stage. Two current sources are used that provide currents $i_1$ and $i_2$, respectively. As the currents may be different, the current $i_b = i_1 - i_2$ flows into the base of NPN or PNP. If $i_b$ is positive, it enters the base of NPN, otherwise it flows out of the base of PNP.

In practical applications, such an ideal push-pull will produce signal distortions when the load current approaches zero because the transistors enter in the non-linear, low current region. This can be avoided by inserting some voltage difference between the two bases. If so, the switching between NPN and PNP is somewhat smoothed, and the distortion is reduced considerably. Such stage is called *class AB amplifier*.

Figure 2.34 shows an example of the power stage of an AB class amplifier. The two current source are made of two NPN transistors $T_1$ and $T_2$; Transistor $T_2$ has fixed operation point. The voltage divider $R_1 - R_2$ provides the constant base polarization. Diode $D_1$ is used to produce a constant voltage drop of 0.6 V that helps to stabilize the operation of $T_2$. Transistor $T_1$ is used as a controlled current source, see Fig. 2.32. The diode $D_2$ and resistance $R_3$ produce a small voltage difference between the bases of the power transistors NPN and PNP in idle state. This is done to provide a small current flow $i_{b1}$, $i_{b2}$ that maintains the power transistor in the lower part of their active linear operation region. Resistances $R_5$ and $R_6$ help to stabilize the idle operation point of NPN-PNP. They work in a way similar to the resistance $R_e$ of the enhanced amplifier of Fig. 2.3. However, these resistances should be very small compared to the load resistance. Note that the load current must flow through $R_5$ or $R_6$. To avoid power loses, the $R_5$ and $R_6$ must have resistance of a fraction of Ohm.

Observe that the power part NPN-PNP can be interpreted as two followers, the PNP viewed upside down.

Let us comment some design hints. Suppose that you want to design an amplifier that can produce up to 20 W of audio signal on a loudspeaker of 8 Ω. This means that the corresponding effective AC voltage $V_{sp}$ on the speaker must be $V_{sp} = \sqrt{8 \times 20} = 12.6$ V.

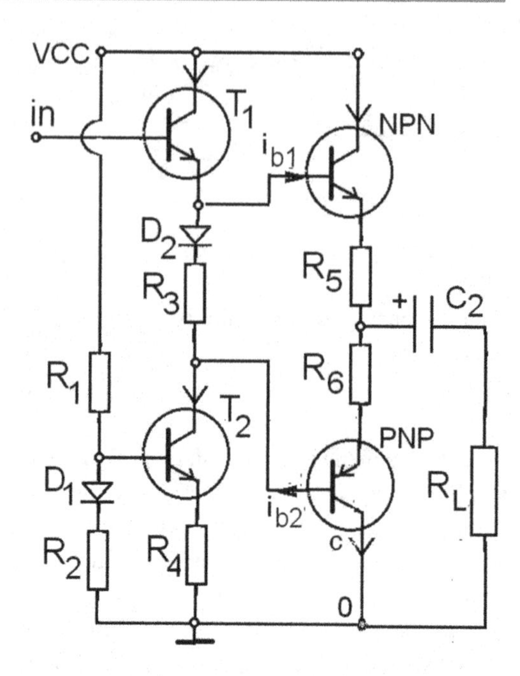


**Fig. 2.34** Push-pull' with controlling circuit

So, the corresponding amplitude of $V_{sp}$ must be equal to $V_s = V_{sp}\sqrt{2} = 17.82$ V. This means that the supply VCC voltage must be at least 35.63 V.

As the efficiency of the amplifier is less than 78.5% , the supplied power must be greater than 25.5 W. So, you must have the power supply circuit of 25.5 W and 35.6 V, or more.

Now, select the transistors. Both NPN and PNP must be power transistors that support the voltage of 37 V. The maximal current $i_max$ can be calculated from the load power. We have $(i_{max}/\sqrt{2}) \times V_{sp} = 20$, so $i_{max} = 1.141 \times 20/17.82 = 1.28$ Ampers. Looking at the catalogs of power transistors, select the complementary pair NPN/PNP with these (or mas stronger) parameters.

Look at the controlling circuit $T_1 - T_2$. Both transistors must support the voltage of 35.5 V. We have selected the power transistors, so we know the $\beta$ of NPN/PNP. Suppose $\beta = 40$. Thus, the maximal current needed to control the complementary pair is $i_max/\beta =$ 32 milliampers. These parameters must be used to select the transistors $T_1$ and $T_2$.

The DC component of the input signal at point "in" must be slightly greater than VCC/2 plus 1.2 V. The voltage drop on $D_2$ and $R_3$ must also be greater than 1.2 V.

Finally, select the (unique) capacitance $C_2$. We are processing an audio signal, so the lowest frequency that must pass to the load is $f_L = 20$ Hz. This means that the cutoff frequency of the circuit $C_2 R_L$ must be less than 20. The capacitance of 1000 μF provides the $f_L = 19.9$ Hz, that satisfies our needs. This may be an electrolytic capacitance, with negative terminal connected to the load.

### 2.9.2 Thermal Instability

In the push-pull, even with ideal components, there is always more than 21.5% of power lost on the output stage. Moreover, some transistors may change parameters with temperature changes. If the transistor $\beta$ grows with growing temperature, then the transistor current increases, and even more heat is dissipated. This results in thermal instability, and can destroy the transistor. A *heat sink* or *heat dissipator* plate is necessary, attached to the transistors. These problems may be avoided by certain enhancement it the final stage, resistors $R_5$, $R_6$ and $R_3$ of Fig. 2.34. Increasing resistors $R_5$, $R_6$ we can improve stability. However, note that the load current must pass through these resistors, so we may lose an important part of the output power. These resistors normally are less then 0.5 Ohm. Other possibility is to use a thermal resistor with negative temperature coefficient (NTC) in parallel with $R_3$. The resistor must be attached to the heat sink to detect the growing temperature. If this occurs, the resistance between collectors of $T_1$ and $T_2$ decreases with growing temperature, and less current is provided to the bases of power transistors. In commercial integrated power amplifiers, this part of the circuit is always somewhat complicated. It can decrease the thermal instability, and may also shut off the circuit when the permissible temperature is exceeded.

### 2.9.3 Amplifier with Feedback

Figure 2.35 shows the push-pull amplifier with voltage amplification input stage. The differential amplifier is used because of its stability and the possibility to use two input points. The input signal is applied to the transistor $T_3$, and the amplified signal is taken from the collector of $T_4$ (non-inverting). this signal is passed to the control circuit of the push-pull stage, described in Sect. 2.9.1.

The bold wire is the feedback connection. The output voltage is passed through the resistance $R_{14}$ to the differential amplifier stage. The signal on the collector of $T_4$ inverts the signal. Note that the transistors $T_1$ and NPN do not invert the signal, so the feedback is negative. The differential amplifier has a great gain. Consequently, the total voltage amplification between points "in" and "out" is defined by the voltage divider $R_14$, $R_b$, where $R_b$ is the resistance of the parallel connection of $R_12$ and $R_13$. So,

$$K = V_{out}/V_{in} = \frac{R_b + R_{14}}{R_b},$$

where

$$R_b = \frac{R_{12}R_{13}}{R_{12} + R_{13}}$$

The gain of an amplifier with feedback has been discussed in Sect. 2.3.

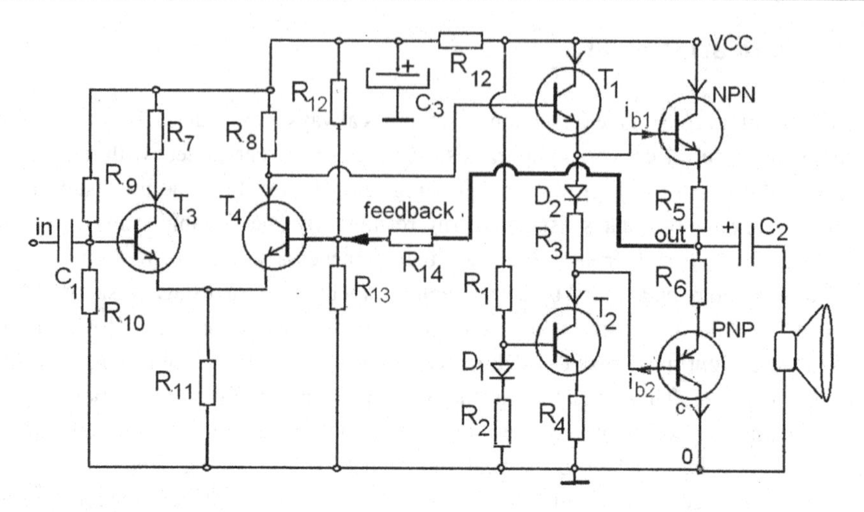


**Fig. 2.35** Audio amplifier

Note the presence of the resistance $R_{12}$ and capacitance $C_3$ in the power supply connection. Remember that any real power supply you can use is not ideal and has certain internal resistance. This means that the strong current fluctuations at the final power stage may affect the voltage VCC. The $R_{12}$ and capacitance $C_3$ prevent such fluctuations to affect the pre-amplification input stage. A reasonable values are 100 Ω and 100 μFarad, respectively.

## 2.10 Class C Amplifiers

In the amplifier of class C we do not care about the linearity and signal distortion. The output stage works outside the active region of the transistor characteristic, below the cut-off point. So, in the idle run no current flows through the power transistors and no power is consumed. Figure 2.36 shows the form of the sinusoidal input and the corresponding output voltage.

The class C amplifiers work normally with signals of high, fixed frequency. The applications are: RF (radio frequency) oscillators, RF amplifier, FM transmitters, high frequency repeaters, tuned amplifiers etc. They have little application in the audio and low-frequency signal processing.

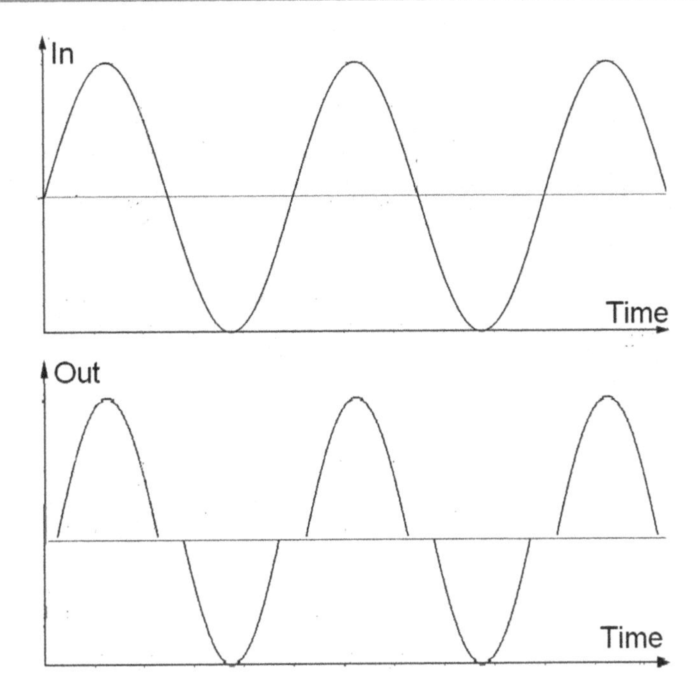


**Fig. 2.36** Class C amplifier

## 2.11 Class D Amplifiers

Recall that the efficiency of the B-class power amplifier is equal to 78.5%. So, if we want to have 60 W of power on the loudspeaker, then approximately 16.4 is lost on the power transistors. The problem is not the cost the additional consumption of power. The main problem is what to do with this 16.4 W of power. The energy must be taken out of the transistors to avoid thermal instability and eventual destruction of the power stage of the amplifier. This is done by the radiators and cooling fans. Anyway, the problem of the dissipated energy is important and sometimes difficult to solve.

The question is: why this dissipated power appears? Looking at the wave forms of the B-class amplifier, we can see that nearly all time, the transistor has both voltage and current different from zero. This produces the unwanted energy. The idea of the D-call amplifier is that power transistor never support voltage and current simultaneously. This occurs, when the transistor operates as a switch. If it is closed, it has no current, and if it is open, the current may be great, but the voltage on the device is nearly null. The D-class amplifier produces a high frequency rectangular signal that is modulated by the pulse duration. Figure 2.37 shows an example of such high frequency signal, and its average, low frequency value.

To obtain modulated rectangular signal, we can compare a high frequency triangular wave with the input signal. What goes out of the comparator device, are the desired modulated

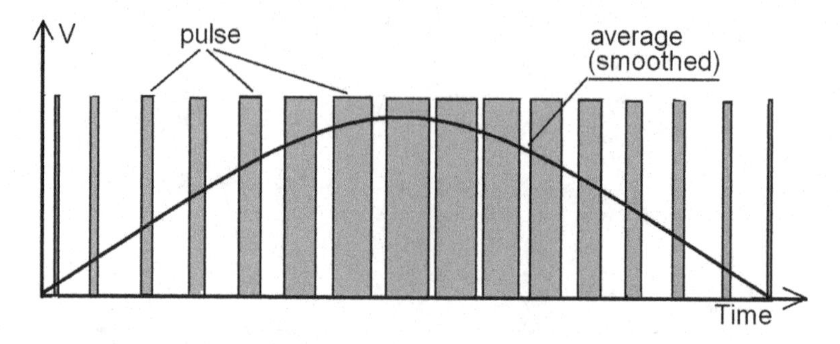


**Fig. 2.37** Modulation by pulse duration

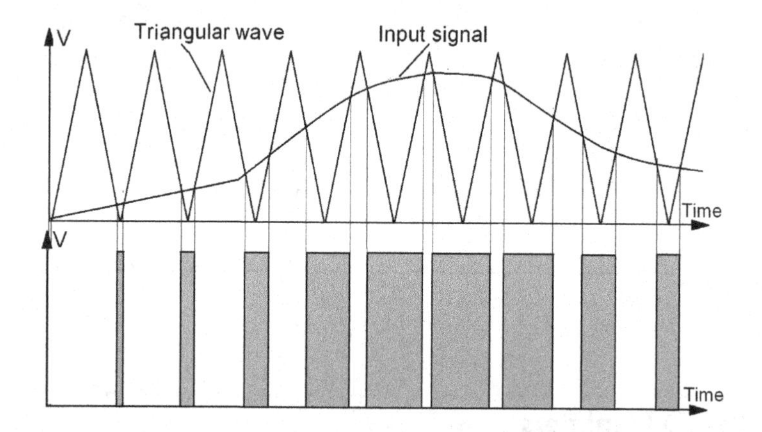


**Fig. 2.38** Obtaining modulated signal

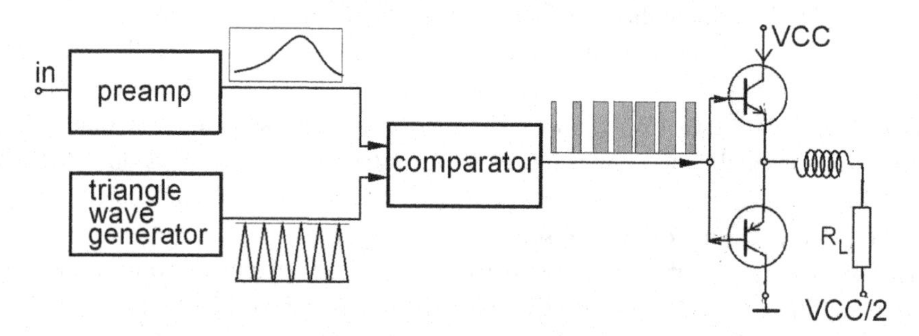


**Fig. 2.39** Class D amplifier

rectangular pulses, as shown in Fig. 2.38. The block scheme of a D-class amplifier is shown in Fig. 2.39.

In Fig. 2.39, we can see a scheme of a C-class amplifier. The amplified acoustic-frequency signal enters to a preamplifier. The comparator compares it with a high frequency triangular wave, and produces the modulated sequence of rectangular pulses. These pulses are applied to the power transistors that work as switching devices. Finally, the amplified pulses are sent to the load. The inductance L has big impedance for high frequencies, so it eliminates them and allows the averaged signal to pass to the load. Note that the transistors work as switches. Consequently, no power is (theoretically) dissipated on them, and the efficiency of the amplifier is approximately 100%.

Figure 2.39 shows only the main idea of the circuit. Here, we use BJT transistors, to compare this circuit with the B-class amplifier discussed earlier. In commercial applications, rather CMOS switching transistors are used.

The parts of the amplifier of Fig. 2.39 can be built using operational amplifiers. Figure 2.40 shows the triangular wave generator, based on operational amplifiers.

The amplifier A generates the rectangular wave. It has a positive feedback from the output x. The voltage on x is passed to the positive input through the resistance $R_2$. Suppose that the amplifier has reached its maximal value VCC. This positive value, while passed to the positive input, maintains this maximal output. However, there is another, negative feedback formed by $R_1$, and the series connection of $R_2$, $R_3$ and $C_1$. This feedback makes the negative input voltage grow with some inertia caused by the capacitance. When the negative input voltage u becomes greater than the voltage v on the positive input, then the amplifier switches to the negative saturation state with output equal to -VCC. The positive feedback maintains $R_2$ maintains this state, until the voltage u becomes low enough the switch the amplifier again. This results in a series of rectangular pulses.

The other amplifier B, is configured as integrator (see Chap. 3). As the integral of a constant value is a ramp, the resulting signal has the form of triangular wave.

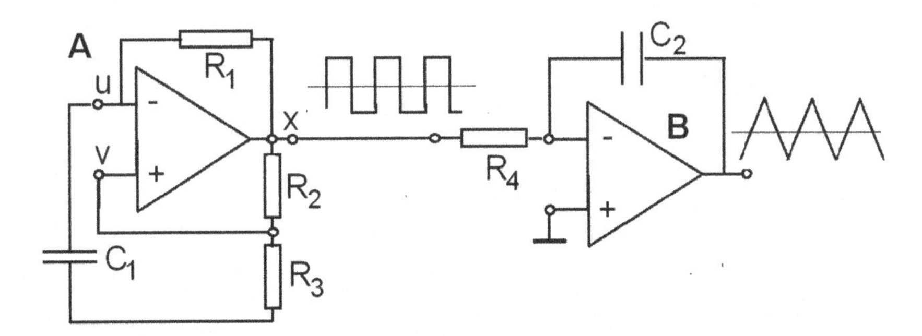


**Fig. 2.40** Triangular wave generator

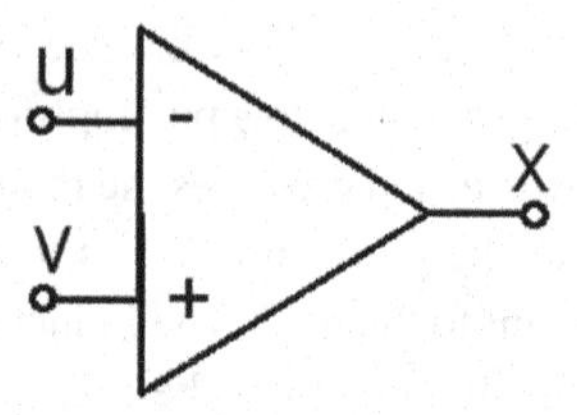


**Fig. 2.41** Comparator

The comparator of Fig. 2.39 can also be built using an op-amp. In fact, an operational amplifier without feedback works as comparator. Due to the huge amplification, it switches between maximal and minimal output voltage, depending on the voltage difference between the two input terminals (Fig. 2.41).

# 3 Filters

## 3.1 Dynamics of Electronic Circuits

Before following with the dynamical characteristics of circuits, recall some facts of the Fourier and Laplace transforms. First of all, note the following:

* Here, the only use of the Fourier approach is to show the spectrum of signals, using the concept of *impedance*.

* The Laplace transform may appear to be a complicated mathematical tool. However, in fact, this tool is used to simplify, and not to complicate the circuit analysis. In the following, we use the most elemental properties of the transform, treating the Laplace complex variable $s$ just as the differentiation operator.

### 3.1.1 Fourier Series

First, we repeat here some useful properties of complex numbers, as given in Sect. 1.6.1.

$$\begin{cases} j^2 = -1 & A \text{ (j - imaginary number unity)} \\ 1/j = -j & B \\ |a + jb| = \sqrt{(a^2 + b^2)} & C \text{ (Absolute value)} \\ \angle(a + jb) = arctan(b/a) & D \text{ (Angle)} \\ e^{j\varphi} = cos(\varphi) + jsin(\varphi) & E \text{ (Euler's representation)} \end{cases} \tag{3.1}$$

Jean-Baptiste Joseph Fourier (1768–1830) developed the theory of the transformation that permits to express a periodic function as a series of sinusoidal terms.

© The Author(s), under exclusive license to Springer Nature Switzerland AG 2023

S. Raczynski, *How Circuits Work*, Synthesis Lectures on Engineering, Science, and Technology, https://doi.org/10.1007/978-3-031-34934-8_3

Consider a function

$$x(t) = x(t + T)\ \forall\ t, \tag{3.2}$$

where $T$ is a constant. The above function is called *periodic function*.

Many of the signals that we analyze or generate are periodic. Most of the musical instruments generate periodic functions. The low frequency signals that come from our heart, and the sound that generate our vocal cords are periodic.

The periodic sinusoidal (one frequency) signal can be express in the following forms:

$$\begin{cases} x(t) = Asin(\omega t + \varphi) & A \\ x(t) = a\,cos(\omega t) + b\,sin(\omega t) & B \\ x(t) = a\cos(\omega t) + jb\,sin(\omega t) & C \\ x(t) = Ae^{j(\omega t + \phi)} & D \end{cases} \tag{3.3}$$

where $\omega$ is the angular frequency [radians/second], $A$ is the signal amplitude and $\phi$ is the phase shift. Note also that $A = \sqrt{(a^2 + b^2)}$, and $\omega = 2\pi f = 2\pi/T$, where $f$ is the frequency [Hz] and $T$ is the function period.

A periodic, integrable function $x(t)$ can be expressed by the *series of Fourier*, as follows.

$$x(t) = \frac{a_o}{2} + \sum_{1}^{\infty}(a_n cos\omega t + b_n sin(\omega t)) \tag{3.4}$$

where

$$\begin{cases} a_o = 2/T \int_{-T/2}^{T/2} x(t)dt \\ a_n = 2/T \int_{-T/2}^{T/2} x(t)cos(\omega t)dt \\ b_n = 2/T \int_{-T/2}^{T/2} x(t)sin(\omega t)dt \end{cases} \tag{3.5}$$

In the complex form, the Fourier series is as follows.

$$\begin{cases} x(t) = \sum_{n=-\infty}^{\infty} c_n e^{j\omega t} \\ c_n = \frac{1}{T}\int_{-T/2}^{T/2} x(t)e^{-jn\omega t}dt \end{cases} \tag{3.6}$$

The Fouries series tells us what is the content of the harmonics $f, 2f, 3f, \ldots$ in the given periodic signal.

### 3.1.2 Laplace Transform

The transform of Laplace of an integrable function $x(t)$ is defined as follows.

$$\mathcal{L}\{x\}(s) = \int_0^{\infty} x(t)e{-}stdt \tag{3.7}$$

The transform $\mathcal{L}$ is a function of a complex variable $s$. Given a Laplace transform, we can retrieve the original function of time $x(t)$ using the corresponding inverse transform formula. However, we will not display this formula because in engineering, the users never calculate inverse transform and do not calculate the integral of Eq. (3.7).

Both transform and its inverse have been calculated for hundreds of most used functions and can be found in Laplace transform (L-transform) tables.

Note the basic properties of the transform.

$$\begin{cases} \mathcal{L}\{Kx\} = K\mathcal{L}\{x\} & A \text{ Linearity} \\ \mathcal{L}\{ax + by\} = a\mathcal{L}\{x\} + b\mathcal{L}\{x\} & B \text{ Linearity} \\ \mathcal{L}\{\frac{dx}{dt}\} = s\mathcal{L}\{x\} & C \\ \mathcal{L}\{\int x\} = \frac{1}{s}\mathcal{L}\{x\} & D \end{cases} \tag{3.8}$$

In the following, we will denote $\mathcal{L}\{x\}(s) = x(s)$.

Let recall that the function "$\delta$ of Dirac is defined as follows.

$$\begin{cases} \delta(t) \equiv 0 \ \forall t \neq 0 \\ \delta(t) = \infty \text{ for } t = 0 \\ \int_{-\infty}^{\infty} \delta(t)dt = 1 \end{cases} \tag{3.9}$$

The *unit step* $1(t)$ is a function equal to one for all $T \geq 0$, zero for $T < 0$.

We have $1(s) = 1/s$ and $\delta(s) = 1$.

The Eq. (3.8C) means that "s" can be treated as the differentiation operator. The integration operator is 1/s.

### 3.1.3 Transfer Function

Consider the differential equation with initial condition x(0) = 0, $u(t)$—given function of time.

$$a\frac{d^2x}{dt^2} + b\frac{dx}{dt} + x(t) = u(t) + d\frac{du}{dt} \tag{3.10}$$

Applying the L-transform to both sides of (6.1) we obtain

$$as^2x(s) + bsx(s) + x(s) = u(s) + dsu(s)$$

This means that the solution in terms of L-transform is

$$x(s) = u(s)\frac{1 + ds}{as^2 + bs + 1}$$

Let define

$$G(s) = \frac{x(s)}{u(s)} = \frac{1 + ds}{as^2 + bs + 1} \tag{3.11}$$

The function $G(s)$ defined by (6.1) describes the dynamics of a device with the input signal $u(t)$ and output $x(t)$. The above function $G(s)$ is the *transfer function* of the device.

The transfer function is the property of the corresponding device. It does not depend on input and output signals. From the control theory it is known that given a transfer function, we can get the system frequency response, simply replacing the variable $s$ by $j\omega$.

Recall that the impedance of the capacitance is $1/(Cj\omega)$. The corresponding impedance in L-transform domain is equal to $Z(s) = 1/(Cs)$. The impedance of the inductor is equal to $Z_L = Ljs$.

## 3.2 Passive Filters

Passive filters are those circuits that can change the content of harmonics in the signal. These devices do not include any amplification elements like transistors or operational amplifiers.

### 3.2.1 RC Circuit

Consider the RC circuit of Fig. 3.1

This is a voltage divider as shown at the right side of the figure. So,

$$x(s) = u(s)\frac{1/(Cs)}{R + 1/(Cs)} = u(s)\frac{1}{RCs + 1} \tag{3.12}$$

In terms of frequency we have

$$\frac{x(j\omega)}{u(j\omega)} = \frac{1}{RCj\omega + 1} \tag{3.13}$$

if $\omega$ approaches zero then the gain of the circuit approaches one. If $\omega$ tends to infinity, the gain of the circuit tends to zero. Thus, the circuit is a *low pass filter*. The absolute value of the ratio $x(j\omega)/u(j\omega)$ is the gain of the circuit for angular frequency $\omega$.

Figure 3.2 shows the spectrum of this circuit. The horizontal scale is logarithmic, from $\omega = 0.01$ to 100 (angular frequency). Horizontal scale is linear in dB (deci-Bell), so, in fact it is also logarithmic. Recall that the gain in dB is equal to $20log_10(K)$, where K is the absolute gain (out/in).

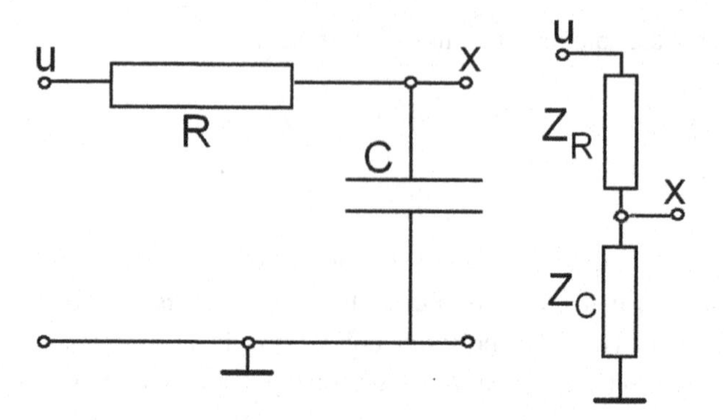


**Fig. 3.1** The RC low-pass filter

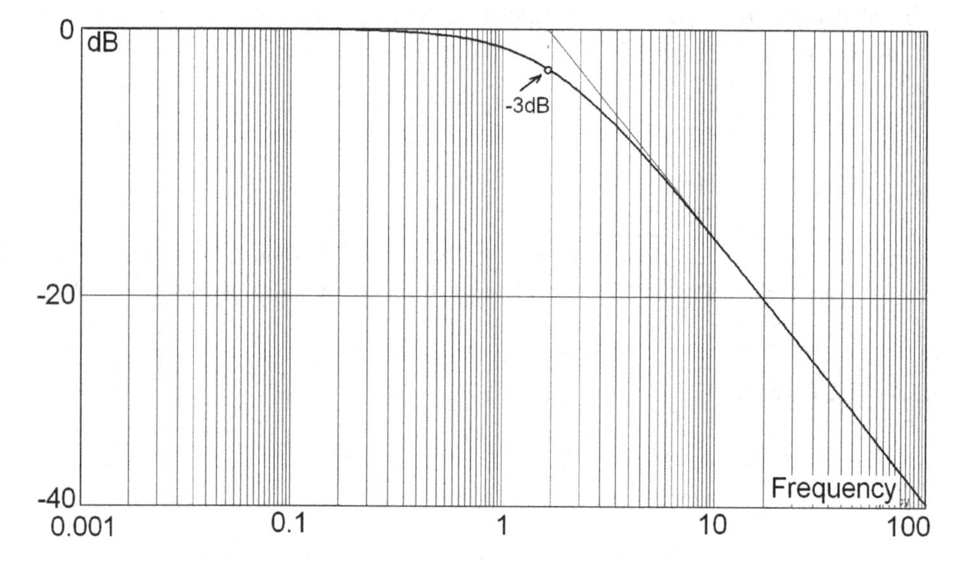


**Fig. 3.2** Frequency response of a RC low-pass filter

It can be seen how the filter eliminates high frequencies. The slope of the characteristic, for high frequencies is equal to $-20$ dB per decade. For the low- and high-pass filters, the important parameter is the *cutofff frequency*. For low-pass filter, this is the frequency for which the gain drops to $-3$ dB with respect to the very low frequencies. If we use a resistance of 1 M$\Omega$ and capacitance of 1 $\mu$Farad, then the cutoff frequency is equal to 0.15915 Hz. The formula for the cutoff frequency is $f_c = 1/(2\pi RC)$. For the cutoff frequency the absolute gain drops to 0.707 of its maximal value.

Figure 3.3 shows the RC high-pass filter.

The formula for the cutoff frequency of the RC high-pass filer is the same as for the low-pass. The difference is that the high-pass filer cuts off low frequencies (Fig. 3.4).

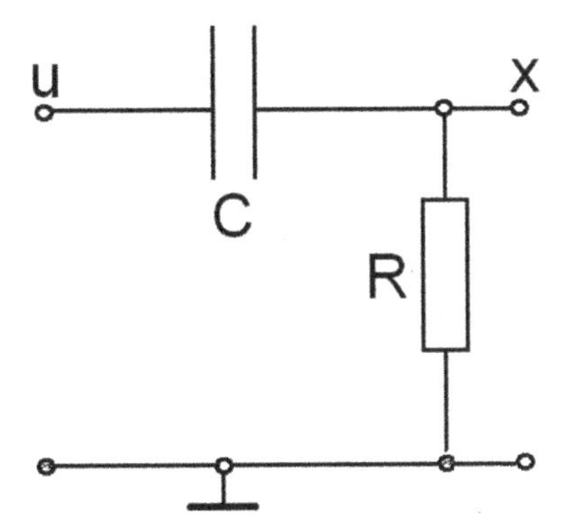


**Fig. 3.3** RC high-pass filter

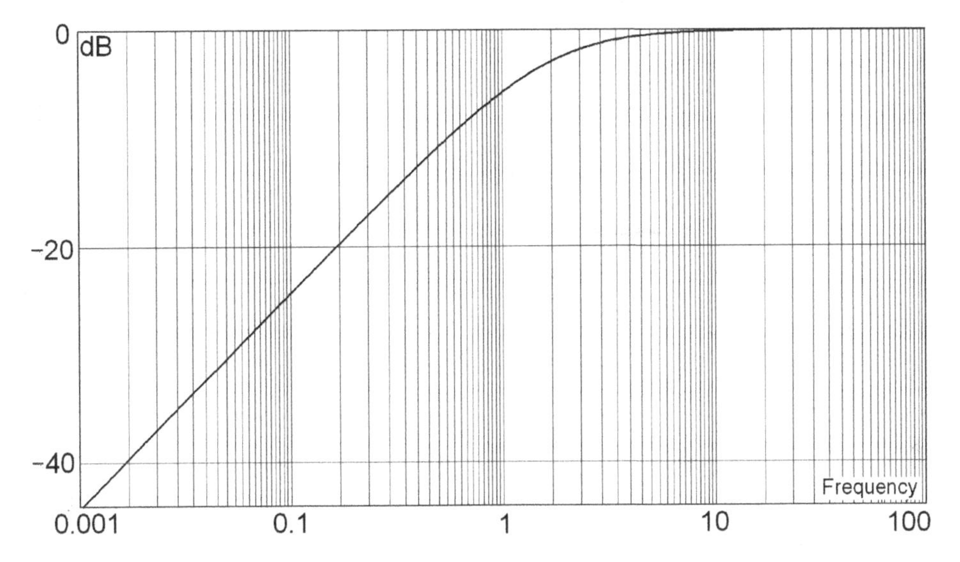


**Fig. 3.4** Frequency response of the RC high-pass filter

### 3.2.2 RLC Circuit

There may be several configurations of the RLC circuit. Let see the circuit of Fig. 3.5.

Following the same procedure as for the RC filter, we obtain

$$x(j\omega) = u(j\omega)\frac{R}{1/(j\omega C) + Lj\omega + R} \tag{3.14}$$

Observe that if $\omega = \sqrt{1/LC}$ then $1/(j\omega C) + Lj\omega = 0$. For this frequency, the gain $x(j\omega)/u(j\omega)$ of the circuit is equal to one. For all other frequencies, the gain in lower. Thus, we have a *band-pass filter*. The band frequency (maximal gain) is $\omega = \sqrt{1/LC}$.

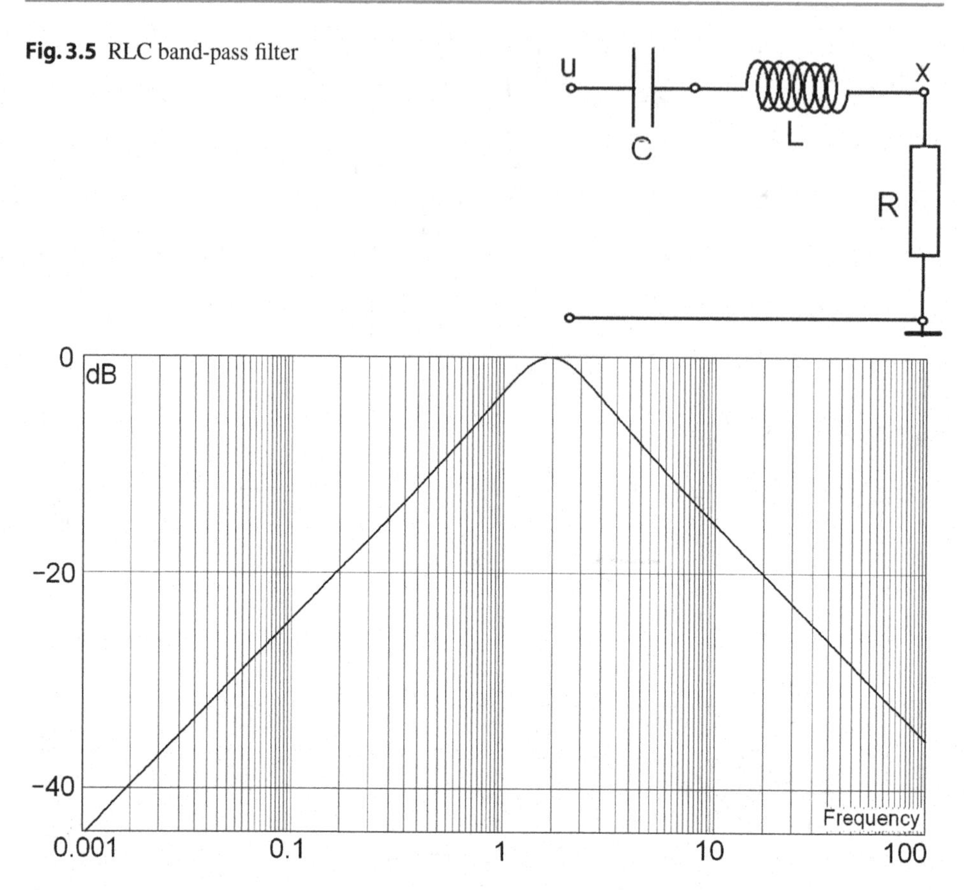


**Fig. 3.5** RLC band-pass filter

**Fig. 3.6** RLC band-pass filter, frequency response

Figure 3.6 shows the spectrum of this device, for LC = 1, R = 1.

Another RLC filter is shown in Fig. 3.7. This is a frequency eliminator. The impedance of the parallel connection LC becomes equal to infinity for the resonance frequency, and the circuit gain becomes equal to zero. As the plot shows the gain in dB, the gain for this frequency drops the minus infinity that is difficult to plot (see Fig. 3.8).

If it is not required that the gain for the resonance frequency becomes zero, then the circuit of Fig. 3.9 can be used. The additional resistance $R_2$ prevents the gain to approach zero, and makes the gain for the eliminated frequency equal to $R/(R + R_2)$.

### 3.2.3 Second-Order RC Filter

Figure 3.10 shows a low-pass second order RC filter. This is **not just a serial connection** of two RC filers of Fig. 3.2 because $R_2 + 1/C_2 s$ represents a complex load for the first stage.

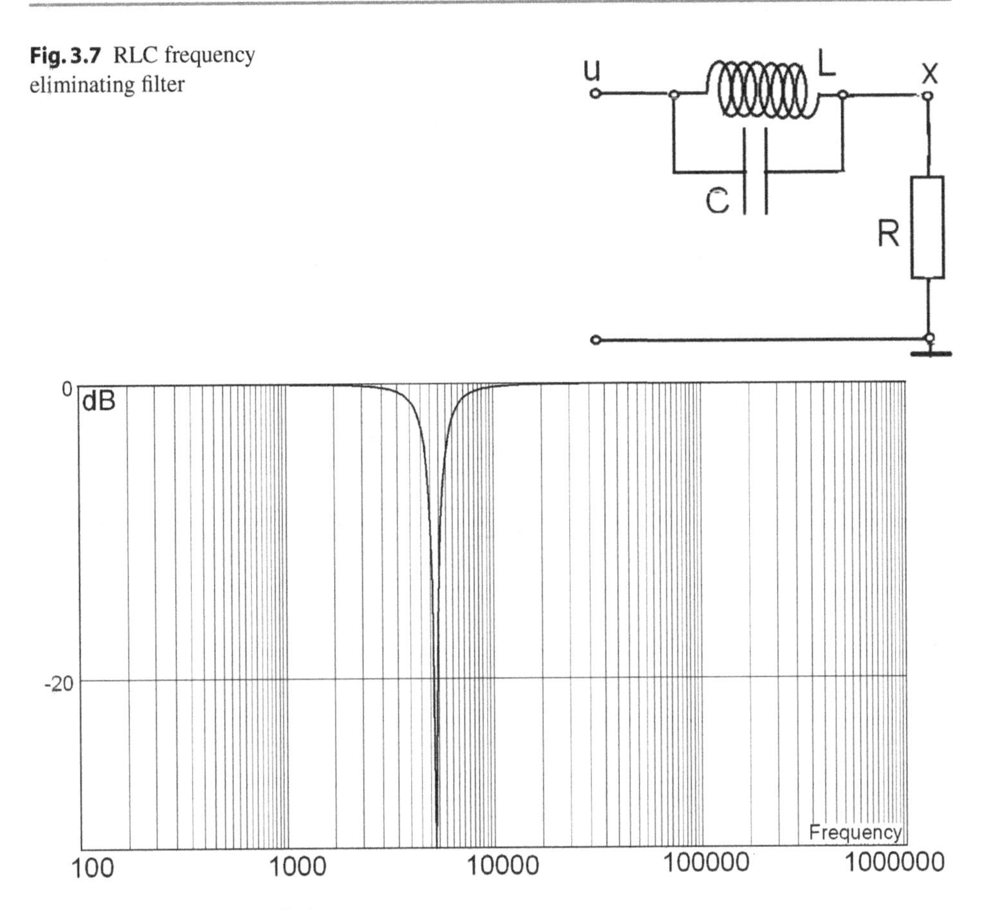


**Fig. 3.7** RLC frequency eliminating filter

**Fig. 3.8** RLC frequency eliminator, spectrum

The transfer function of this circuit is as follows

$$G(s) = \frac{x(s)}{u(s)} = \frac{1}{R_1 R_2 C_1 C_2 s^2 + s(R_1 C_1 + C_2(R_1 + R_2)) + 1} \tag{3.15}$$

Note that the slope of the characteristic for high frequencies is $-40$ dB per decade.

The frequency response of this filter with $C_1 = C_2 = 10\,\text{nF}$ and $R_1 = R_2 = 10\,\text{k}\Omega$ is shown in Fig. 3.11.

### 3.2.4 Wien Bridge

Another passive pass-band filter is shown in Fig. 3.12. It is a second order filter, based on resistencies and capacitances. It can be used as a filter, but its mas common application is in the oscillators.

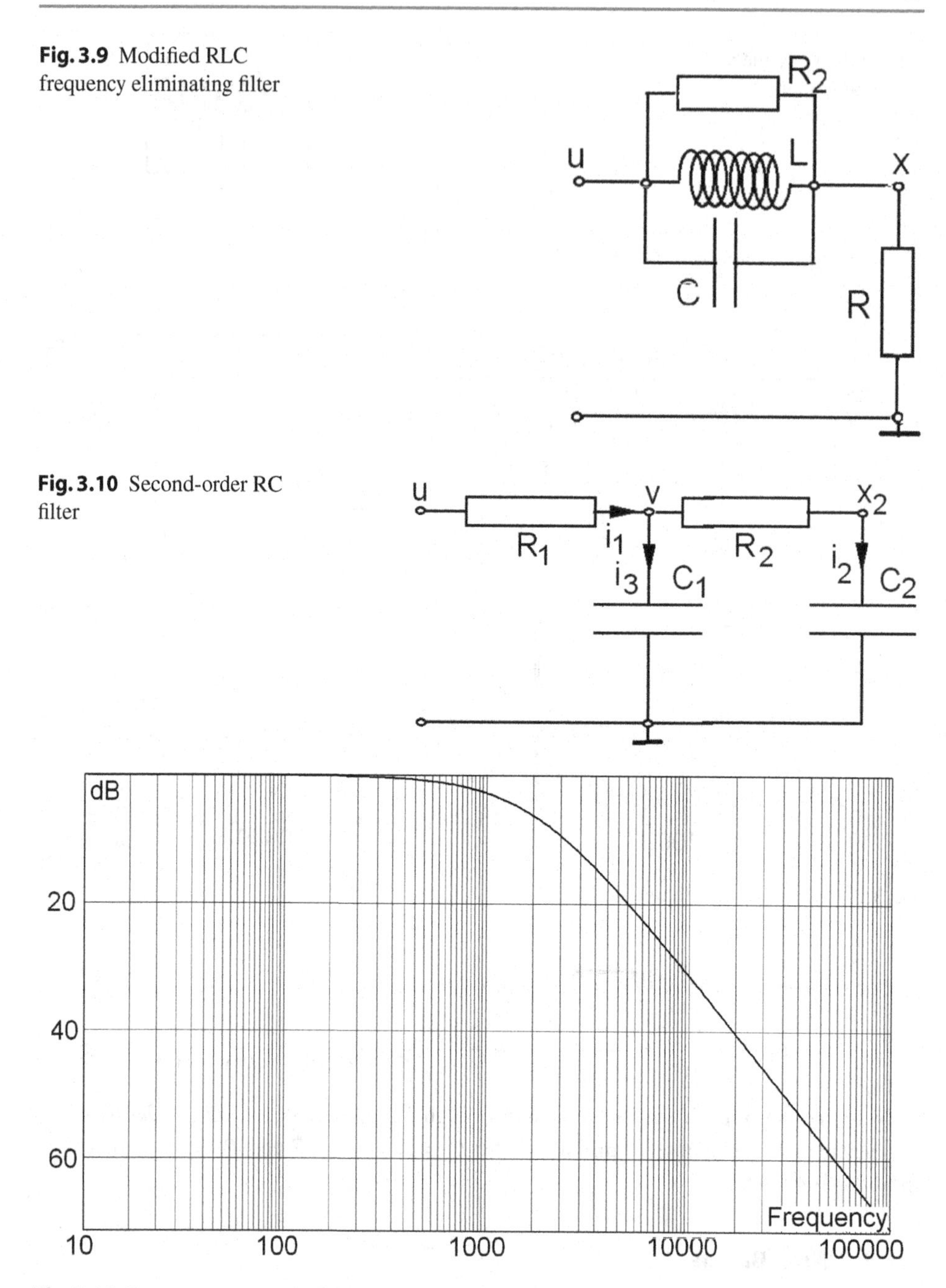


**Fig. 3.9** Modified RLC frequency eliminating filter

**Fig. 3.10** Second-order RC filter

**Fig. 3.11** Frequency response of the RC low-pass second order filter

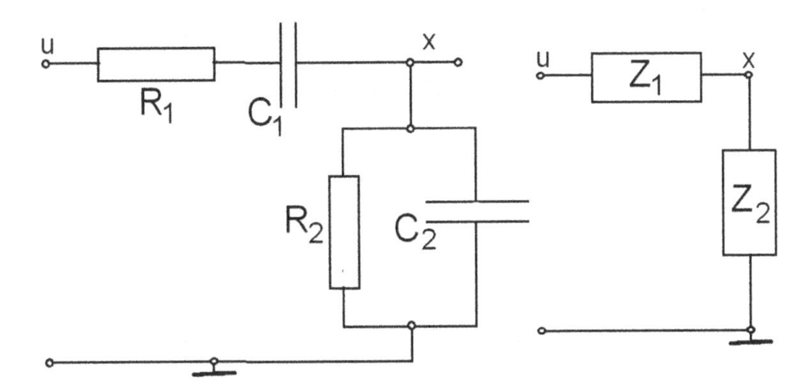


**Fig. 3.12** Wien bridge circuit

The right part of Fig. 3.12 shows the equivalent voltage divider, with two impedances $Z_1$ and $Z_2$. We have

$$Z_1 = R_1 + \frac{1}{C_1 s}$$

$$Z_2 = \frac{R_2/(C_2 s)}{R_2 + 1/(C_2 s)}$$

So, from the voltage divider $Z_1 - Z_2$ podemos calculate voltage $x$:

$$x(s) = u(s)\frac{Z_2}{Z_1 + Z_2} \tag{3.16}$$

Finally, after substituting expressions for $Z_1$, $Z_2$ and reordering we get the transfer function

$$G(s) = \frac{x(s)}{u(s)} = \frac{R_2 C_1 s}{R_1 C_1 R_2 C_2 s^2 + (R_1 C_1 + R_2 C_2 + R_2 C_1)s + 1} \tag{3.17}$$

The band frequency of this circuit is

$$f = \frac{1}{2\pi\sqrt{R_1 R_2 . C_1, C_2}}$$

An example of the frequency response (in dB) of Wien bridge, with $R_1 = R_2 = R_3 = 1000\,\Omega$ and $C_1 = C_2 = C_3 = 100\,\text{nF}$, is shown in Fig. 3.13. Figure 3.14 depicts the same response as the absolute gain [out/in]. The corresponding phase shift can be seen at Fig. 3.15. It can be seen that the frequency response of this filter is similar as that of the RLC band-pass filter (Fig. 3.6).

An advantage of the Wien bridge filter is that it does not use inductances.

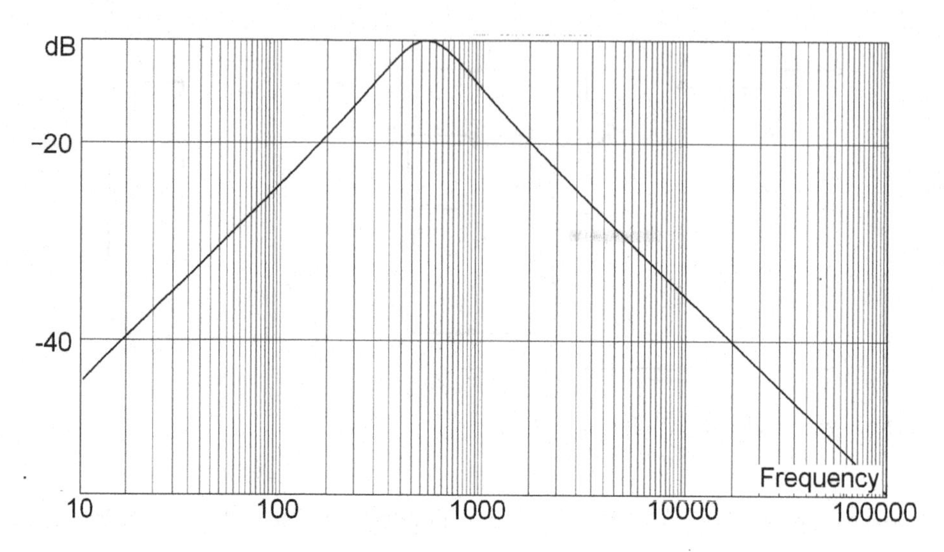


**Fig. 3.13** Puente Wien: frequency response [dB]

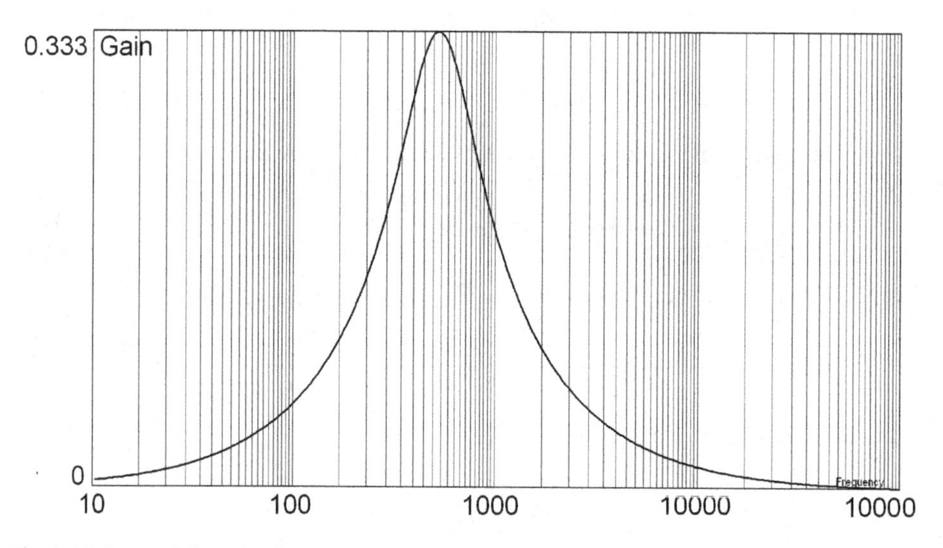


**Fig. 3.14** Puente Wien: absolute gain

### 3.2.5 Twin T Filter

Figure 3.16 shows an useful passive, frequency eliminating circuit "Double T" or "TT-circuit". The transfer function is as follows.

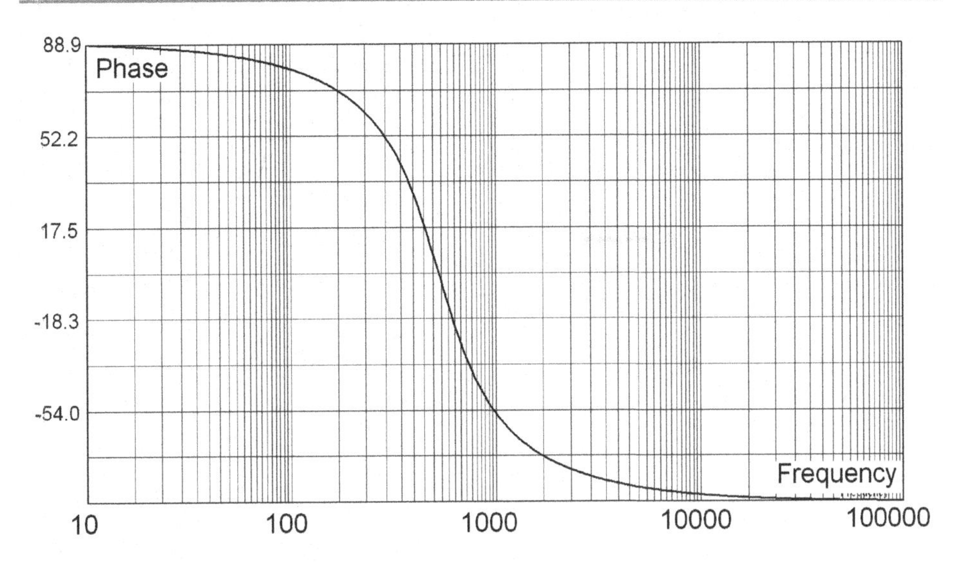


**Fig. 3.15** Puente Wien: phase shift

$$\begin{cases} G(s) = \dfrac{x(s)}{u(s)} = \\ \dfrac{s^3 + s^2\left(\frac{1}{C_1R_1} + \frac{1}{C_1R_2}\right) + s\left(\frac{1}{C_1C_3R_1R_2} + \frac{1}{C_1C_2R_1R_2}\right) + \frac{1}{C_1C_2C_3R_1R_2R_3}}{s^3 + As^2 + Bs + C} \\ \text{where} \\ A = \dfrac{1}{C_1R_1} + \dfrac{1}{C_1R_2} + \dfrac{1}{C_2R_2} + \dfrac{1}{C_2R_3} + \dfrac{1}{C_3R_12} \\ B = \dfrac{1}{C_1C_3R_1R_2} + \dfrac{1}{C_1C_2R_1R_2} + \dfrac{1}{C_1C_2R_1R_3} + \dfrac{1}{C_1C_2R_3R_4} + \dfrac{1}{C_2C_3R_2R_3} \\ C = \dfrac{1}{C_1C_2C_3R_1R_2R_3} \end{cases} \tag{3.18}$$

Figure 3.17 shows the frequency response for TT-circuit with $C_1 = C_2 = C_3 = 20\,\text{nF}$ and $R_1 = R_2 = R_3 = 10\,\text{k}\Omega$. In Fig. 3.18 we can see the corresponding phase shift.

The TT circuit has good frequency response that does not drop to zero. It is useful in audio applications.

### 3.2.6 Treble and Bass Control

A simple, but commonly used circuit for audio treble and bass control is shown in Fig. 3.19.

The part $R_1 - C_1 - C_2 - R_2$ forms a low-pass filter. If the potenciometer $P_1$ is in position a, then the low frequencies filtered by $R_1$, $C_1$, $C_2$ pass through $R_3$ to the output. If it is in position b, then the low frequencies decrement. Potenciometer $P_2$ controls

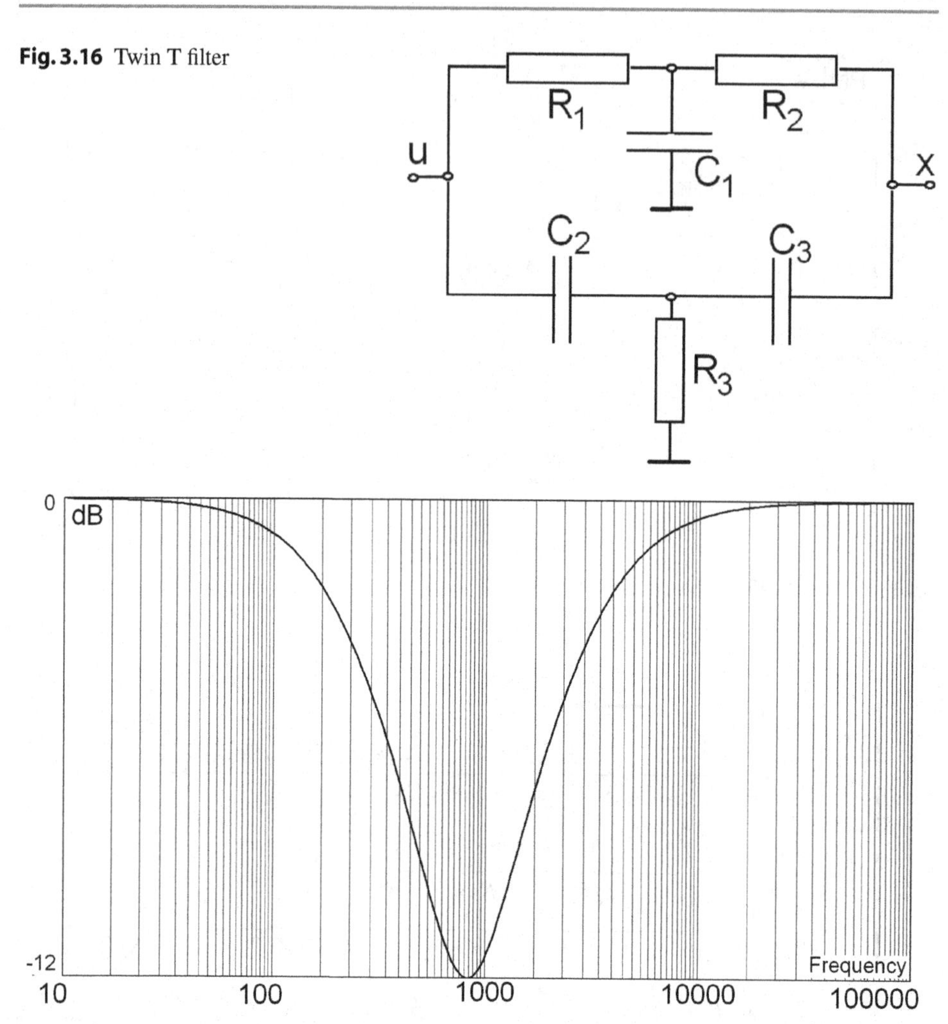


**Fig. 3.16** Twin T filter

**Fig. 3.17** TT-circuit: frequency response

the treble (high frequencies). If it is in position c, then high frequencies pass through $C_4$ to the output. If $P_2$ is in low position d, then the high frequencies are eliminated because $P_2$ and $C_5$ form a low-pass filter. The recommended values for this circuit are as follows: $R_1 = 3.3\,\text{k}\Omega$, $R_2 = 3.3\,\text{k}\Omega$, $R_3 = 6.8\,\text{k}\Omega$ $P_1 = 100\,\text{k}\Omega$, $P_2 = 100\,\text{k}\Omega$, $C_1 = 0.1\,\mu\text{F}$, $C_2 = 0.1\,\mu\text{F}$, $C_3 = 10\,\mu\text{F}$, $C_4 = 1\,\text{nF}$, $C_5 = 1\,\text{nF}$.

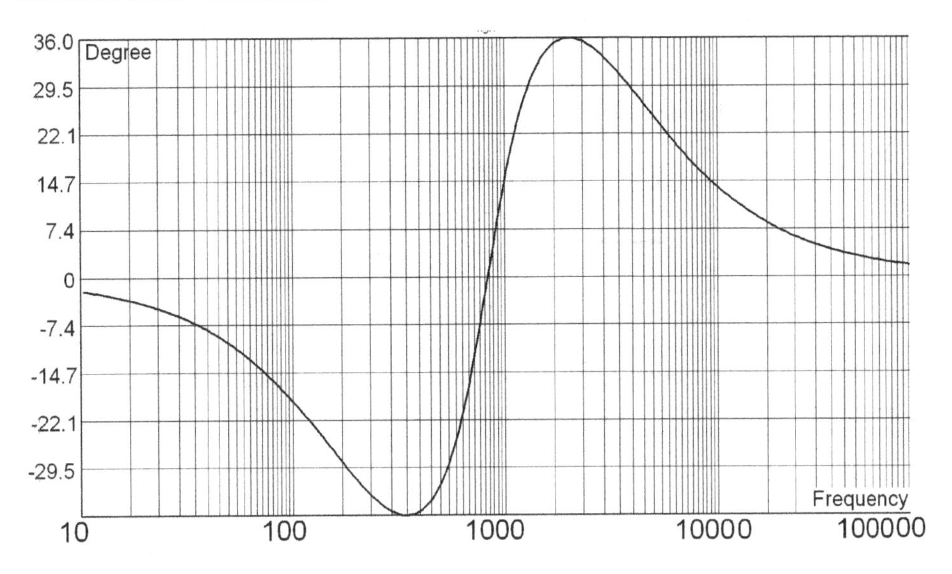


**Fig. 3.18** TT-circuit: phase shift

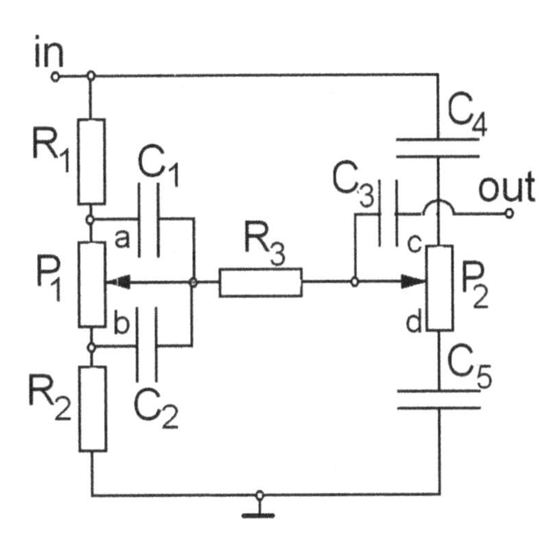


**Fig. 3.19** Treble and bass control

## 3.3 Active Filters, Op-amps with Feedback

The *active filter* is a device that includes one or more active elements that can amplify signals. This can be one or more stage transistor amplifiers, but the most used element is the operational amplifier (op-amp), described in Sect. 2.8. Here, we will use and "ideal" op-amp, supposing that it is a highly linear device with differential input, very high input impedance and very low output impedance. The gain K of the op-amp is supposed to be very high, approaching infinity. The ideal op-amp has no internal inertia or signal delays.

In Sect. 2.8 a, op-amp with negative feedback has been discussed. Now, consider the feedback formed by complex impedances, as in Fig. 3.20.

Note that $i_1 = i_2$ because the op-amp has practically infinite input impedance. From Fig. 3.20 we have

$$\begin{cases} e = u + (x - u)\dfrac{Z_2}{Z_1 + Z_2} \\ x = -Ke = -K\left(u + (x - u)\dfrac{Z_2}{Z_1 + Z_2}\right) \\ x\left(1 + K\dfrac{Z_2}{Z_1 + Z_2}\right) = -Ku\dfrac{Z_1}{Z_1 + Z_2} \end{cases} \tag{3.19}$$

if $K \to \infty$, we can neglect the "1" in the last equations. This means that the total gain of the circuit (the transfer function) is as follows:

$$G(s) = \frac{x(s)}{u(s)} = -\frac{Z_1(s)}{Z_2(s)} \tag{3.20}$$

Note that if $Z_2 = R$ is a real impedance (a resistance), then the transfer function G is the $Z_1$ with coefficient R. For example, if $Z_2 = R$, $Z_1 = 1/(CS)$, then the resulting transfer function is $G(s) = 1/(RCs)$ (an integrator). With $R = 1\,\mathrm{M}\Omega$ and $C = 1\,\mu\mathrm{F}$, we get $G(s) = 1/s$

Figure 3.21 depicts an example of two most used op-amps with feedback. The transfer function of the circuit in part A is

$$G(s) = -\frac{1/(Cs)}{R} = -\frac{1}{RC}\frac{1}{s}$$

This means that the circuit is an integrator with gain coefficient $1/(RC)$. The circuit B is a differentiator, $x(s) = -RCs\,u(s)$. The op-amp of part C is just an amplifier with gain $G = -R_1/R_2$, and the circuit D is a follower where $x = u$ (note that this circuit does not invert the phase of the signal). Further on, we will use simplified symbols, as shown in Fig. 3.22.

Now, let us see an op-amp with negative feedback through a transfer function $H(s)$ shown in Fig. 3.23. From the figure we obtain:

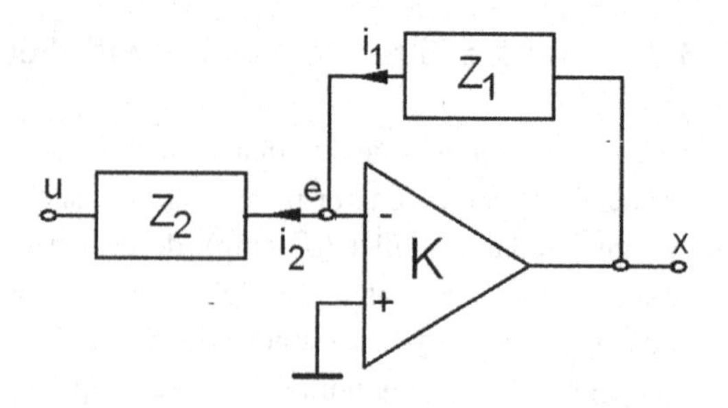


**Fig. 3.20** op-amp with feedback

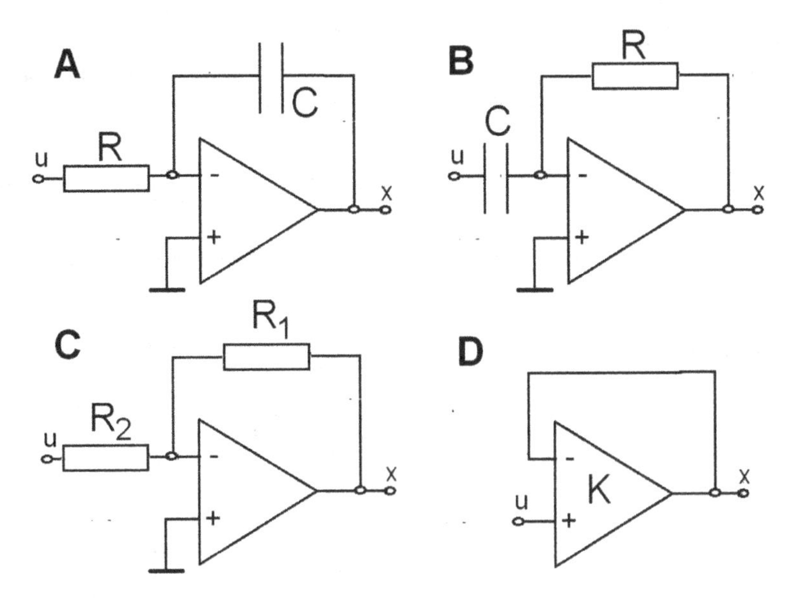


**Fig. 3.21** Integrator and differentiator

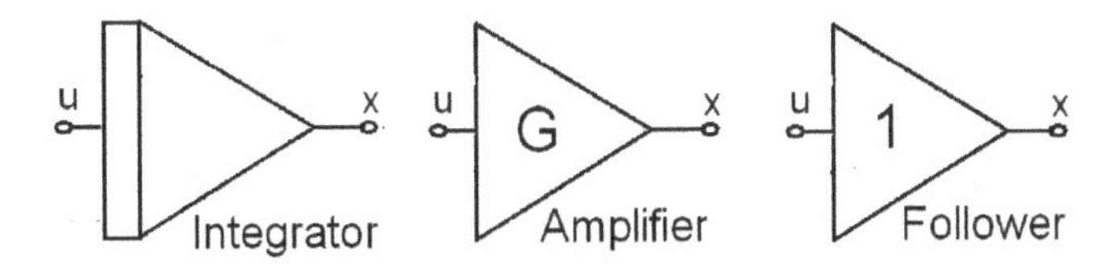


**Fig. 3.22** Simplified symbols

$$\begin{cases} v = x(s)H(s) \\ x(s) = K(u(s) - v(s)) = K(u(s) - x(s)H(s)) \end{cases} \tag{3.21}$$

and, finally, with $K \to \infty$, the transfer function of the circuit is

$$G(s) = \frac{x[s]}{u(s)} = \frac{1}{H(s)} \tag{3.22}$$

In other words, the transfer function is the reciprocal of the feedback function. Here, we use the symbol of the operational amplifier. However, any amplifier can be used, provided it has differential input and big gain, for example a one- or two-stage differential amplifier.

### 3.3.1 The Sumator

In Fig. 3.24 we can see a sumator circuit. If all the resistances are equal to each other, then at the output we get the sum of all input voltages $u_1, u_2, u_2, \ldots, u_n$, with negative sign. The

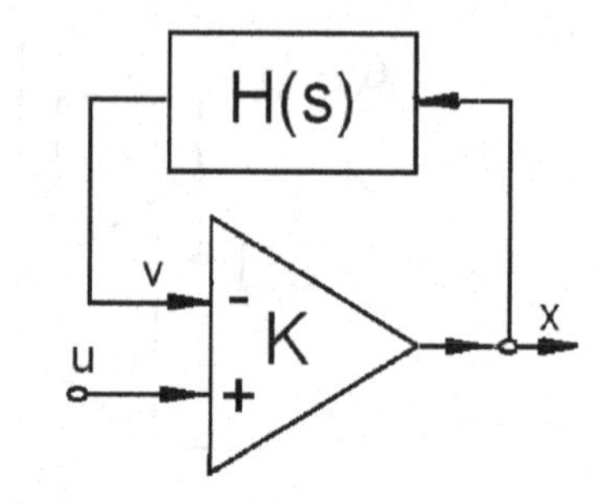


**Fig. 3.23** op-amp with feedback H(s)

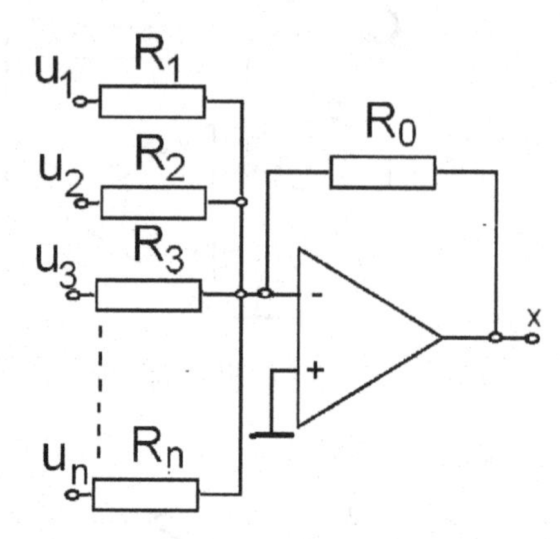


**Fig. 3.24** <the sumator.

resistance $R_0$ may be used to control the overall gain, and the input resistances may control the gains for the corresponding inputs.

### 3.3.2 Treble and Bass Boost

Consider the circuit of Fig. 3.25. In the feedback we have a circuit with transfer function $H(s)$. Supposed that both $S_1$ and $S_2$ are open. So, $C_1$ can be ignored, and the current i is as follows:

$$i = \frac{x}{R_1 + R_2 + R_3 + 1/(C_2 s)}$$

So,

$$v = i R_3 = R_3 \frac{x}{R_1 + R_2 + R_3 + 1/(C_2 s)}$$

and the feedback $H(s)$ is

$$\begin{cases} H(s) = \dfrac{v(s)}{x(s)} = R_3 \dfrac{x}{R_1 + R_2 + R_3 + 1/(C_2 s)} \\ H(s) = \dfrac{R_3 C_2 s}{(R_1 + R_2 + R_3) C_2 s + 1} \end{cases} \tag{3.23}$$

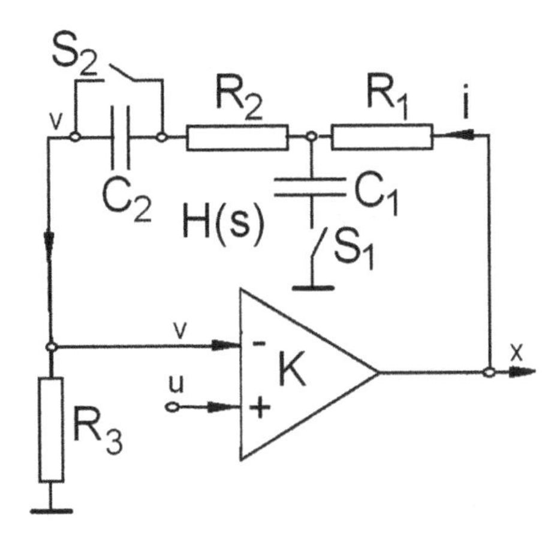


**Fig. 3.25** Treble and bass boost

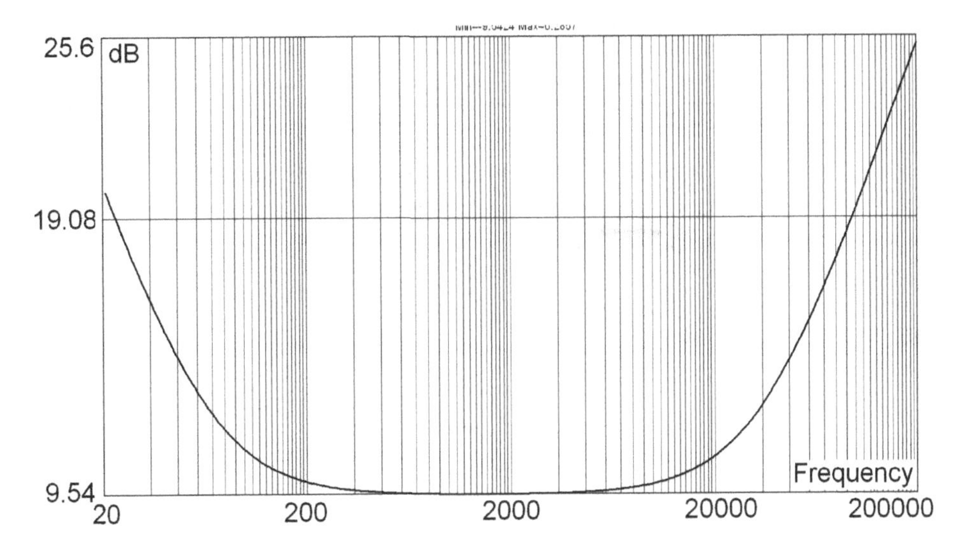


**Fig. 3.26** Treble and bass boost

This means that the total gain is

$$G(s) = \frac{x(s)}{u(s)} = \frac{1}{H(s)} = \frac{(R_1 + R_2 + R_3)C_2 s + 1}{R_3 C_2 s} \tag{3.24}$$

Figure 3.26 shows the frequency response with $S_1$ closed and $S_2$ open (both bass and treble boost). The parameters are: $C_1 = 0.2\,\text{nF},\ C_2 = 50\,\text{nF},\ R_1 = R_2 = R_3 = 25\,\text{k}\Omega$

Observe that the characteristic is approximately plain between 250 and 2500Hz, and both bass and treble are amplified considerably.

### 3.3.3 Feedback with Wien Bridge

Consider the circuit of Fig. 3.27.

The transfer function of the Wien bridge has been discussed in Sect. 3.2.4. The transfer function between nodes $x$ and $v$ is as follows.

$$F(s) = \frac{v(s)}{x(s)} = \frac{R_2C_1s}{R_1C_1R_2C_2s^2 + (R_1C_1 + R_2C_2 + R_2C_1)s + 1} \tag{3.25}$$

So, we have

$$\begin{cases} v(s) = x(s)F(s) \\ x(s) = -K(u(s) - v(s)) = -K(u(s) - x(s)F(s)) \\ x(s)(1 + KF(s)) = -Ku(s) \end{cases} \tag{3.26}$$

When $K \to \infty$, from the third equation of (3.26) we obtain

$$G(s) = \frac{x(s)}{u(s)} = \frac{1}{F(s)} = \frac{R_1C_1R_2C_2s^2 + (R_1C_1 + R_2C_2 + R_2C_1)s + 1}{R_2C_1s} \tag{3.27}$$

Let $C_1 = C_2 = 10\,\text{nF},\ R_1 = R_2 = 10\,\text{k}\Omega$. Figure 3.28 shows the frequency response of the circuit. It can be seen that we get a frequency eliminating (notch) circuit, with characteristic that do not go to negative infinity. This can be useful in audio devices.

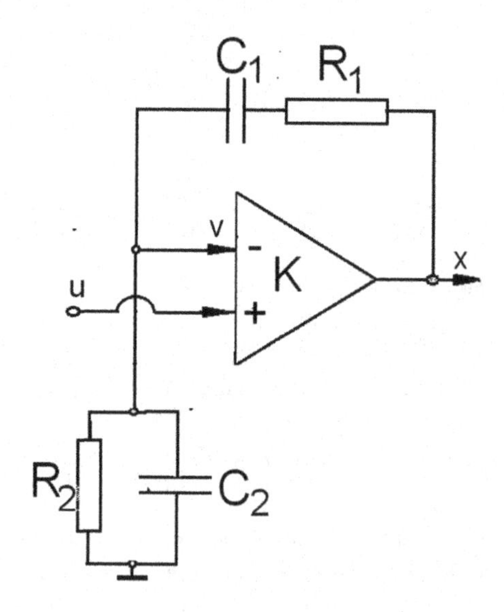


**Fig. 3.27** Feedback with wien bridge

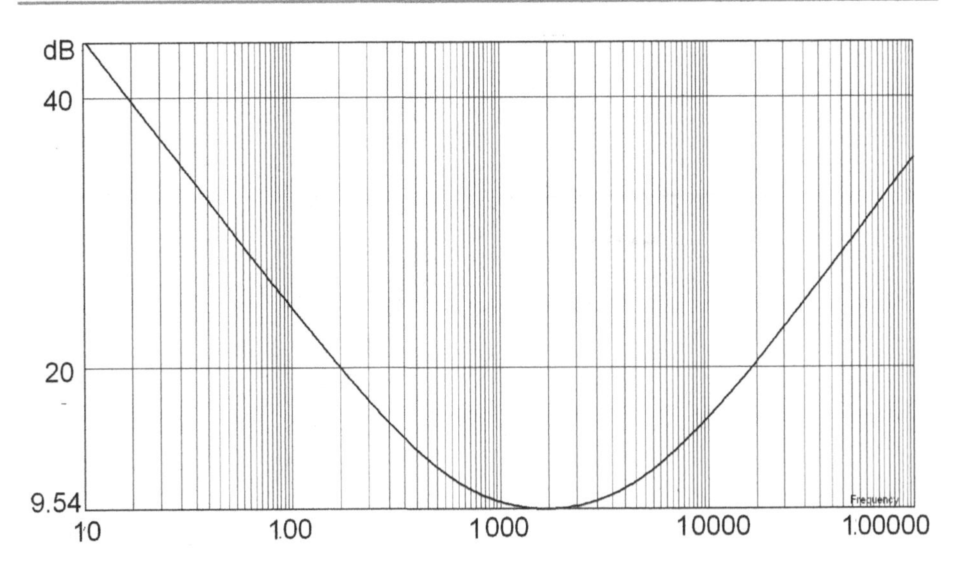


**Fig. 3.28** Feedback with wien bridge—frequency response

## 3.4 Higher Order Filters and Transfer Functions

### 3.4.1 Second Order Filter

Before discussing more filters, let recall the standard forms of a second order transfer function, as follows

$$\begin{cases} G(s) = \dfrac{K\omega_o^2}{s^2 + 2\xi\omega_o s + \omega_o^2} & \text{low pass} \\ G(s) = \dfrac{K2\xi\omega_o s}{s^2 + 2\xi\omega_o s + \omega_o^2} & \text{band pass} \\ G(s) = \dfrac{Ks^2}{s^2 + 2\xi\omega_o s + \omega_o^2} & \text{high pass} \end{cases} \tag{3.28}$$

where $K$ is the overall filter gain. In other forms of these transfer function, the "filer quality" factor $Q$ is used, $Q = 1/(2\xi)$. Note that $\omega_o$ is a fixed parameter of the transfer function,

In the time domain, these objects are described by linear second-order differential equations (see Sect. 3.1.2). For example, the above low-pass filter equation is

$$\frac{d^2x}{dt^2} + 2\xi\omega_o\frac{dx}{dt} + \omega_o^2 x(t) = K\omega_o^2 u(t), \tag{3.29}$$

where $u(t)$ is the input signal, and $x(t)$ is the filter response.

Let us see the properties of the low-pass filter of Eq. (3.28). The parameter $\omega_o$ defines the cut-off angular frequency equal to $2\pi f_o$, where $f_o$ is the frequency in Hz. Parameter $\xi$ defines the dumping. For $\xi < 1$ the time response of the filter is oscillatory, and the filter

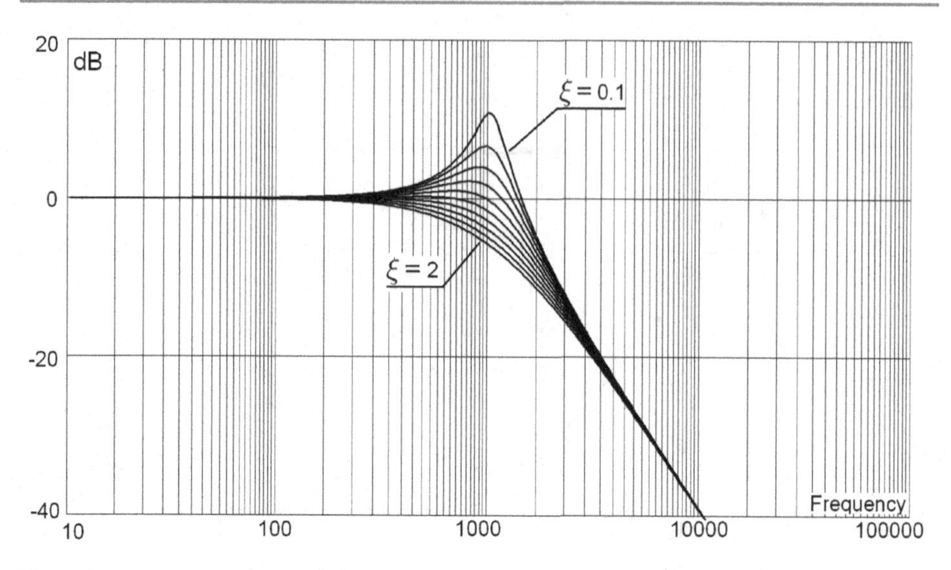


**Fig. 3.29** Low-pass second order filter. Frequency response for various $\xi$

may enter a resonance for frequency $f_o$. This means that the absolute gain for this frequency increases. If $\xi > 1$, the device is overdumped, and no oscillations or resonance can appear.

For example, consider the low-pass filter with the parameter: $\omega = 6,283.18$ rad/sec that corresponds to frequency $f = 1000$ Hz.

Figure 3.29 shows the frequency response of the filter for some different values of $\xi$, between 0.1 and 2.

A good shape of the frequency response can be achieved for dumping lower than 1. In this case, we obtain certain gain increase for frequencies near $f_o$, and then a rapid decrease for higher frequencies, see Figs. 3.30 and 3.31. The slope of the characteristic is equal to $-40$ dB per decade (compared to $-20$ dB per decade for the first order filter).

Figure 3.32 depicts the frequency response of the second order band-pass filter (see Eq. (3.28)), with $\xi = 0.1$. The absolute gain is shown in Fig. 3.33. It can be seen that with low $\xi = 0.1$ we can obtain a band-pass filter with very narrow band width.

We will not discuss here the high-pass filter of Eq. (3.28) because its frequency response is mostly just a horizontal flip of that of the low-pass filter.

### 3.4.2 Second Order Filters with Op-amps

Consider the transfer function of a low-pass filter of Eq. (3.28). This corresponds to the differential equation

$$\begin{cases} \dfrac{d^2x}{dt^2} + a\dfrac{dx}{dt} + bx(t) = c\,u(t) \\ \dfrac{d^2x}{dt^2} = c\,u(t) - a\dfrac{dx}{dt} - bx(t) \end{cases} \tag{3.30}$$

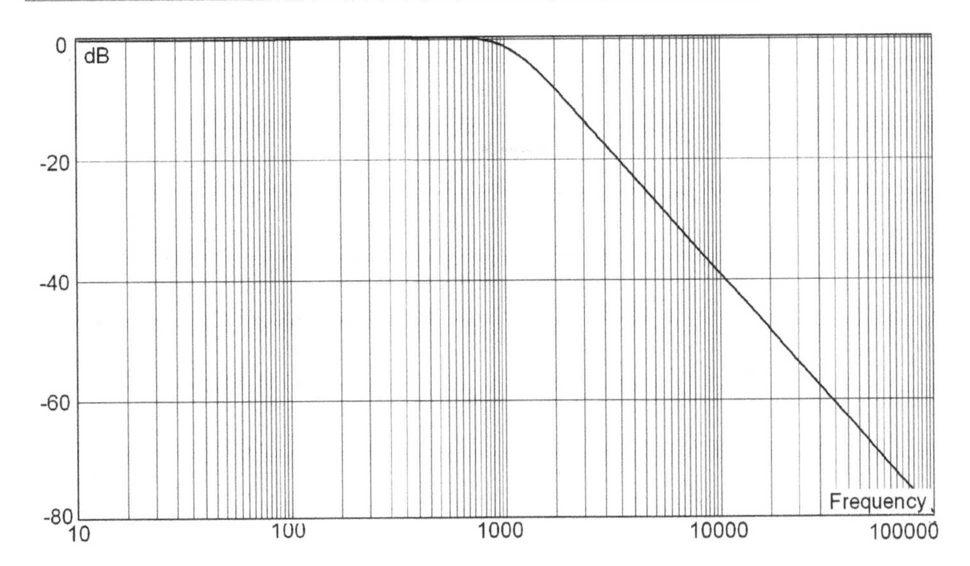


**Fig. 3.30** Low-pass second order filter. Frequency response for $\xi = 0.6$

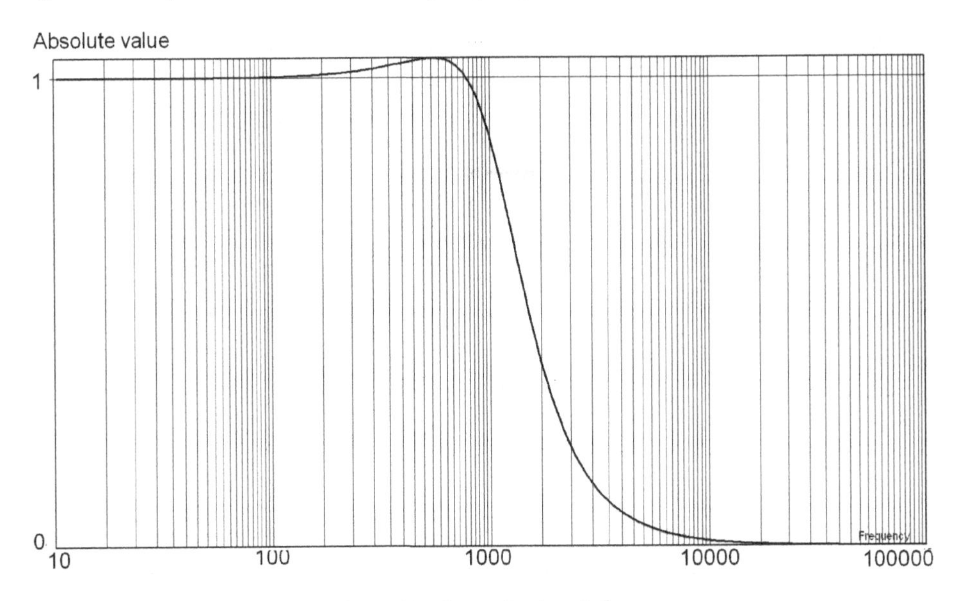


**Fig. 3.31** Low-pass second order filter. Absolute gain, $\xi = 0.6$

Now, look at the circuit of Fig. 3.34. The op-amps marked with "i" are integrators, the element "-1" solo inverts the signal sign. Dot on top means time-differentiation. The op-amp "s" is configured as a sumator (also inverting the signal sign). Note that an integrator with overall gain $-1$ may be done using an op-amp with the capacitance 1 μF in feedback, and 1 MΩ at the input. This kind of circuits are referred to as "state variable models".

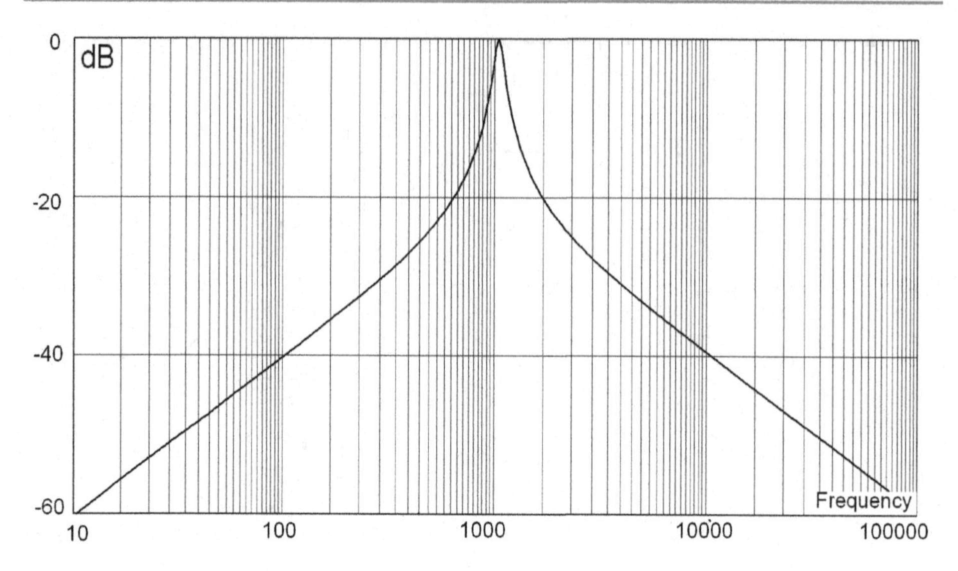


**Fig. 3.32** Band-pass second order filter. Frequency response, $\xi = 0.1$

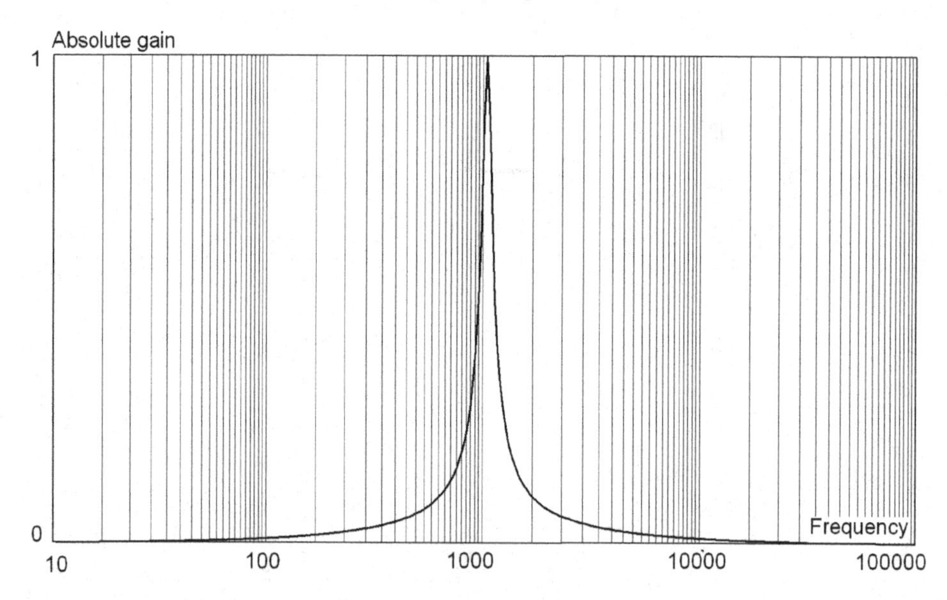


**Fig. 3.33** Band-pass second order filter. Absolute gain, $\xi = 0.1$

Suppose that at the circuit output we have signal $x(t)$. Then, at the input of the last integrator must be $-\dot{x}$, and on the input to the previous integrator we must have $\ddot{x}$. But this point is, at the same time, the output from the sumator. So, $\ddot{x} = u - a\dot{x} - bx(t)$. This means that the signal $x(t)$ satisfies Eq. (3.30).

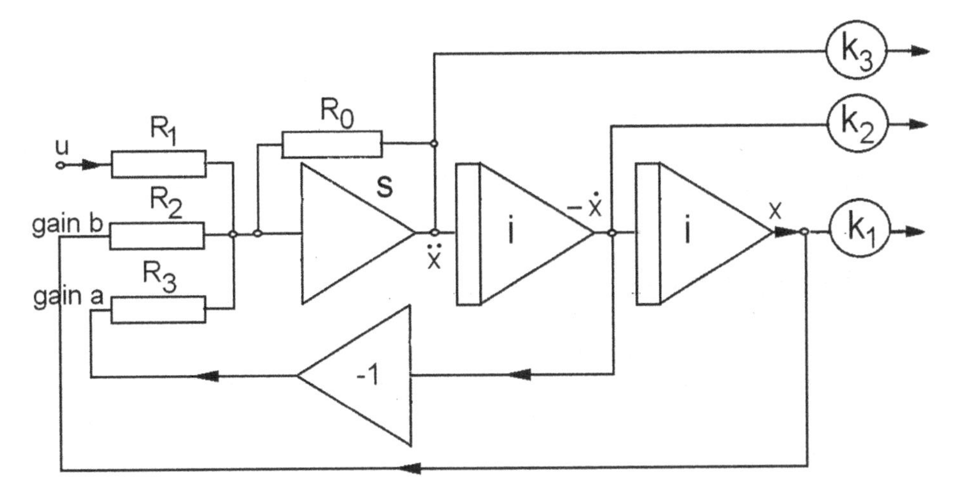


**Fig. 3.34** A second order filter

If we apply any signal $u(t)$ at the input $R_1$, then we have the filter response at $x(t)$. The coefficients $a$ and $b$ can be adjusted by appropriate selection of $R_0$, $R_3$ and $R_2$, respectively.

Theoretically, this model works for any coefficients a and b. However, if the frequency $f_o = \omega_o/(2\pi)$ is equal to 1000 Hz, as in the previous examples, then the coefficients $a = b = \omega^2$ should be equal to 39478417. It is difficult to adjust such big coefficients, selecting corresponding resistances $R_0$, $R_1$ and $R_2$. If we use very small and very big resistances, problems with stability and saturation may appear. The solution is to change the gain of the integrators. For example, if we use the resistance of 10 kΩ and capacitance equal to 0.1 μF, then the integrators produce signals equal to $-1000 \int(input)$ each one. This means that at the input "gain a" should have gain equal to 0.001a, and the gain b should be equal to 0.000001 b. These values are the most appropriate in practical application. It is easy to check that the whole circuit with these parameters will satisfy the same differential equation.

Taking the output signal from $x$, $\dot{x}$ and $\ddot{x}$, with coefficients $k_1$, $k_2$ and $k_3$, we can obtain a band-pass and a high-pass filters.

If we connect three of more integrators in series, then we can construct filters of higher order, with the transfer function as of Eq. (3.31).

$$G(s) = \frac{b_0 + b_1 s + b_2 s^s + b_3 s^3 + \cdots + b_m s^m}{a_0 + a_1 s + a_2 s^2 + a_3 s^3 + \cdots + a_n s^n}, \tag{3.31}$$

where $m \leq n$.

## 3.5 Higher Order Filters

The second order filter can provide frequency rejection with characteristics slope −40 dB per decade. To achieve better frequency elimination, we can use filters of higher order, described by Eq. (3.31).

### 3.5.1 Chebyshev Filter

Note that a Russian mathematician *Pafnuti Lvóvich Chebyshov*, referred also as *Chebyshev* (1821–1894) never invented any filter or other signal-processing device. The name of this kind of filter comes after his works on the theory of polynomials.

By a *polynomial* we mean the expression

$$f(x) = a_n x^n + a_{n-1} x^{n-1} + a_{n-2} x^{n-2} + \cdots + a_0 x, \tag{3.32}$$

where $a_n \neq 0$.

If the independent variable $x$ tends to infinity, the absolute value of the polynomial $f(x)$ may oscillate, but for big $x$ it also must go to infinity. In few words, the *polynomial of Chabyshev* has the coefficients $a_n$ defined in such a way, that $f(x)$ sticks to zero at the interval [0, 1], and then may go to infinity.

The transfer function of the filter we design is given in the form $G(s) = M(s)/N(s)$, where $M$ and $N$ are polynomials. This fact is used to relate the filter design to the polynomials of Chebyshev. The application of Chebyshev polynomials resulted in the filters that have approximately flat frequency response inside the given bandwidth, and strong frequency rejection outside.

A Type I Chebyshev low-pass filter has an all-pole transfer function. A Type II Chebyshev low-pass filter has both poles and zeros. The absolute value of the frequency response in terms of angular frequency $j\omega$ is as follows:

$$|H(j\omega)| = \frac{1}{\sqrt{1 + \varepsilon^2 T^2(\omega/\omega_0)}}, \tag{3.33}$$

where $\varepsilon$ is the *ripple* parameter.

The design procedures and the theory of Chebyshev filters are explained in more details in the specialized literature. However, the practical design is not very complicated. The coefficients of transfer functions have been calculated for various versions of the filters that vary mainly with the ripple parameter, see [5]. The tables of transfer function coefficients and/or poles can be found in many sources. Below is an example of a fragment of such table. These are the denominator polynomials of low-pass filters of order n = 1, 2 and 3, with cutoff frequency is $\omega = 1,\ f = 0.159$ Hz. ripple 0.5 dB.

$$n = 1 : s + 2.8627$$

$$n = 2 : s^2 + 1.4256s + 1.5162$$

$$n = 3 : s^3 + 1.2529s^2 + 1.5349s + 0.715$$

Figure 3.35 shows the frequency response of the 3rd order filter. It may appear similar to the first and second-order filters discussed earlier, but is is quite different: note that the characteristics slope for f > 159 Hz is equal to 60 dB per decade. The filter can be constructed as a state variable mode of order three, like that of Sect. 3.4.2. It can also be created as the serial connection of one stage of the first order, and the other of the second order, with a pair of complex poles.

Figure 3.36 depicts the absolute gain of the filter.

In the tables of Chebyshef transfer functions, the cutoff frequency is normally fixed to $\omega = 1$ [rad/sec]. The filter can be adjusted to any other frequency by multiplying the coefficients as follows:

$$a_k' = a_k \left(\frac{\omega_0}{\omega_1}\right)^k = a_k \left(\frac{\omega_0}{2\pi f_1}\right)^k \tag{3.34}$$

where $\omega_0$ is the frequency of the table (in our case equal to one), and $\omega_1$ and $f_1$ are desired frequencies (angular and Hz, respectively).

In Fig. 3.37 we can see an example of a circuit that may work as the third order Chebyshev filter. A similar, second order filter has been discussed earlier (see Fig. 3.34). The coefficients $a_n$ can be adjusted by the feedback resistances as follows:

$$R_s/R_0 = a_0, \ R_s/R_1 = a_1, \ R_s/R_2 = a_2$$

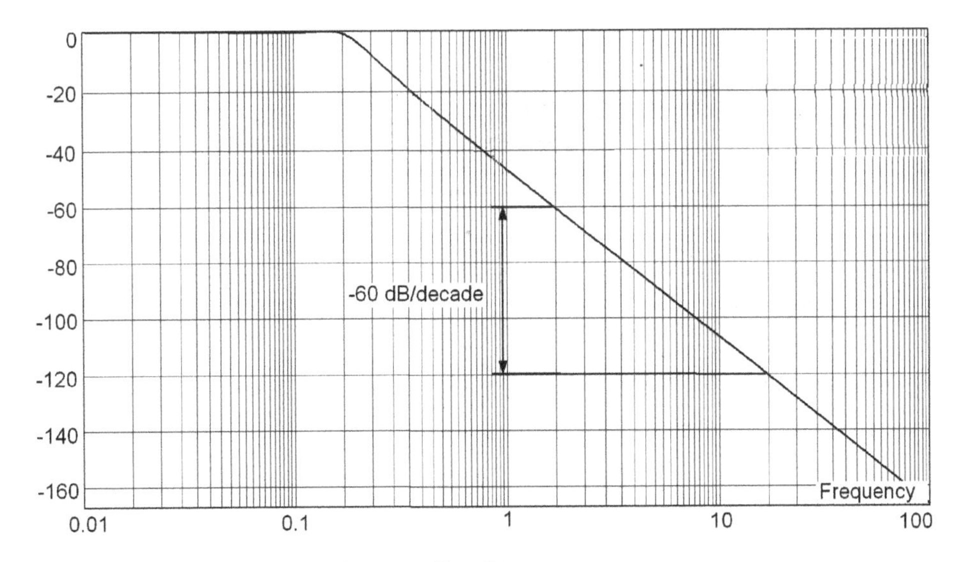


**Fig. 3.35** Third order Chebyshev low-pass filter. Frequency response

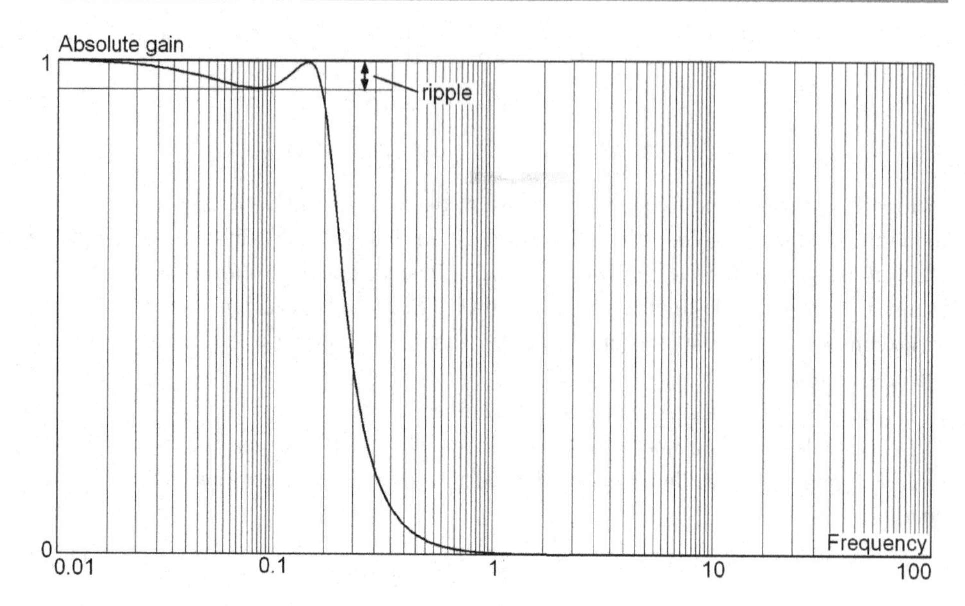


**Fig. 3.36** Third order Chebyshev low-pass filter. Absolute gain

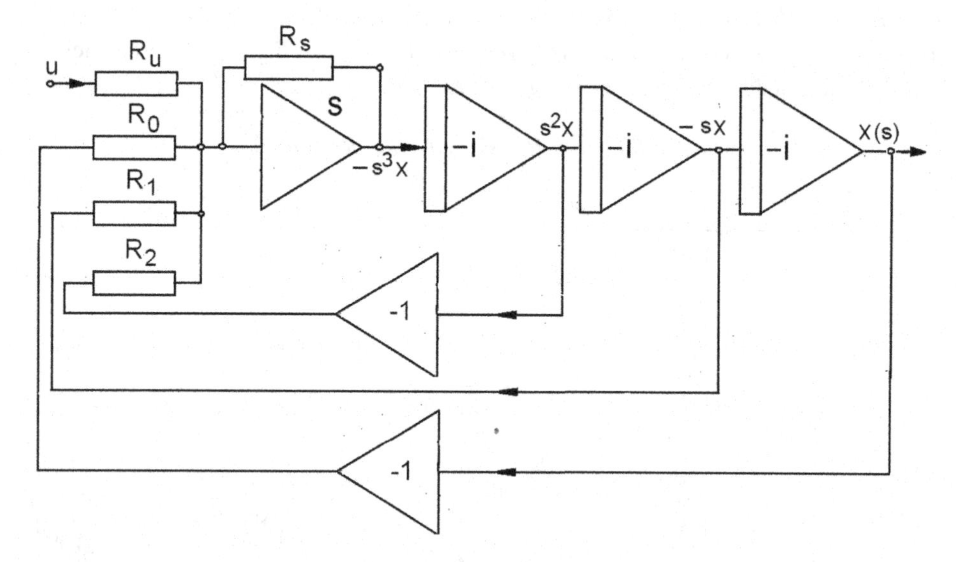


**Fig. 3.37** State variable realization of a third order filter

### 3.5.2 Butterworth Filter

In 1930, the British engineer Stephen Butterworth described the filter that is as flat as possible in the passband.

The transfer function of a Butterworth filter is as follows:

$$G(s) = G_0 \prod_{k=1}^{n} \frac{\omega_0}{s - \omega_0 e^{\frac{j(2k+n-1)\pi}{2n}}} \tag{3.35}$$

Like for the Chebyshev filter, the coefficients of the Butterworth transfer function have been calculated and published in tables. For example, below there are the denominators for the filters of order 1, 2 and 3:

$$n = 1 : s + 1$$
$$n = 2 : s^2 + 1.4142s + 1$$
$$n = 3 : s^3 + 2s^2 + 2s + 1$$

Figure 3.38 shows the frequency response of the third order Butterworth filter for cutoff frequency $\omega = 1,\ f = 0.159\,\text{Hz}$.

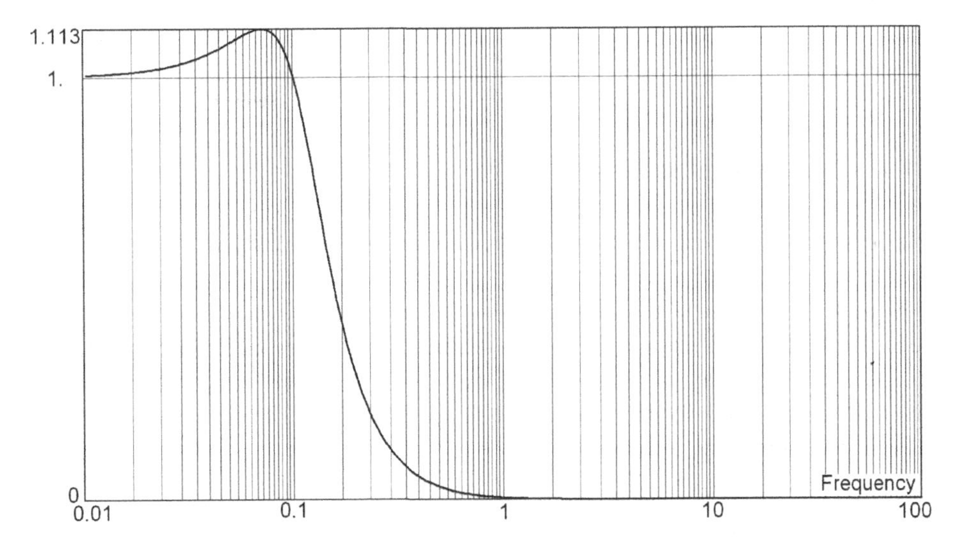


**Fig. 3.38** Third order Butterworth low-pass filter. Absolute gain

# Oscillators, Multivibrators, Triggers and Flip-Flops 4

## 4.1 LC Oscillators

The resonance of the LC circuit is used to construct sinusoidal oscillators with relatively high frequencies. The circuits are quite simple and robust. Normally, a transistor is used as the active element.

As the circuits have a positive feedback, they are always unstable. If they were linear circuits without saturation, the oscillations would go to infinity. In the real circuit, the oscillations grow, until they reach the limit of linear active region of the transistor or the amplifies. Then, the oscillations stabilize. In LC oscillators, the LC circuit eliminates frequencies different from the resonance frequency, so the output signal is approximately sinusoidal, with little distortion.

### 4.1.1 Tuned-Base Oscillator

This circuit, also known as *Armstrong* or *Meissner* oscillator, implements a coupled inductance to make the positive feedback.

Figure 4.1 shows one of the simplest solutions. The transistor works in common emitter mode. The collector current passes through the primary part of a coupled inductance. The voltage of the secondary part is applied to the base. The connections are made to apply the inverted signal to the base that meas the positive feedback. The circuit LC defines the frequency of oscillations, as follows

$$f = \frac{1}{2\pi\sqrt{LC}}$$

© The Author(s), under exclusive license to Springer Nature Switzerland AG 2023

S. Raczynski, *How Circuits Work*, Synthesis Lectures on Engineering, Science, and Technology, https://doi.org/10.1007/978-3-031-34934-8_4

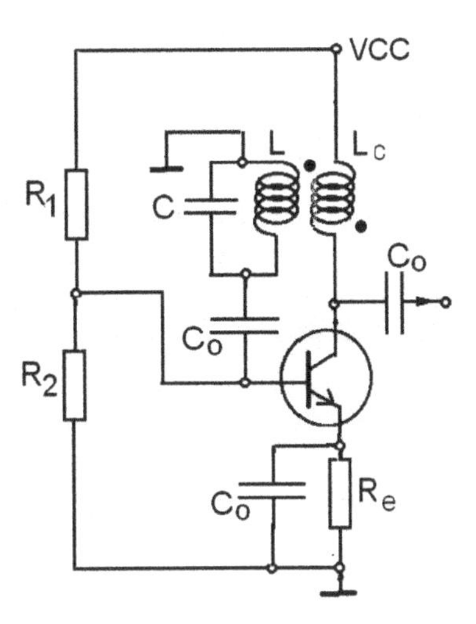

**Fig. 4.1** Basic, tuned-base LC oscillator

The resistances $R_1$ and $R_2$ define the operation point of the transistor, and the resistance $R_e$ is used to improve the stability of the operation point. The capacitances marked as $C_0$ should have big values to reach a small impedance for the generated frequency.

### 4.1.2 Colpitts Oscillator

This oscillator, invented by Edwin Henry Colpitts (1872–1949) is one of the most popular sinusoidal oscillators. It uses an inductance connected to a "divided" capacitance, as shown in Fig. 4.2. The circuit LC that determines the generated frequency consists of the inductance L and two capacitances $C_1$ and $C_2$ connected in series. The equivalent capacitance is as follows:

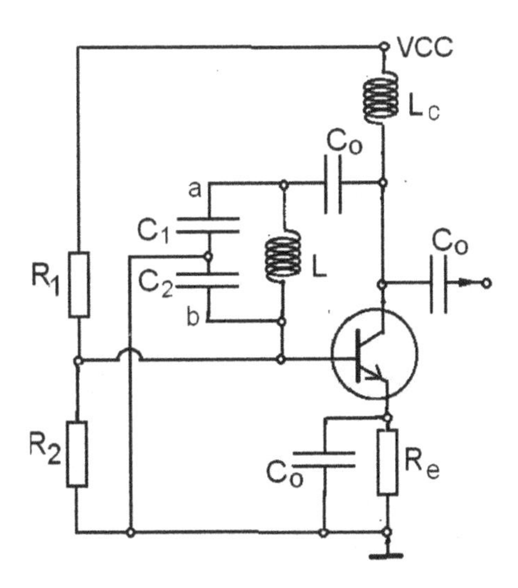

**Fig. 4.2** Colpitts oscillator

$$C = \frac{C_1 C_2}{C_1 + C_2} \tag{4.1}$$

The point of connection between the capacitances is connected to the ground. This makes the phase of the oscillation at point b opposite to the phase at the collector. The other, big inductance $L_c$ is used only to pass the current to the transistor collector. It has no influence on the frequency. As in the tuned-base oscillator, the resistances $R_1$ and $R_2$ define the operation point of the transistor, and the resistance $R_e$ is used to improve the stability of the operation point. The capacitances marked as $C_0$ should have big values to reach a small impedance for the generated frequency.

### 4.1.3 Hartley Oscillator

Figure 4.3 shows the scheme of the Hartley oscillator. Unlike Colpitts, this oscillator uses the LC circuit formed by coupled inductances $L_1$, $L_2$ and $C$. The frequency is

$$f = \frac{1}{2\pi\sqrt{C(L_1 + L_2)}}$$

The point between $L_1$ and $L_2$ is connected to the ground. So, the signals at points a and b have opposite phase that makes the feedback positive. $R_1$ and $R_2$ define the operation point of the transistor, and the resistance $R_e$ is used to improve the stability of the operation point. The capacitances marked as $C_0$ should have big values to reach a small impedance for the generated frequency. The inductance $L_C$ is used only to pass the current to the collector.

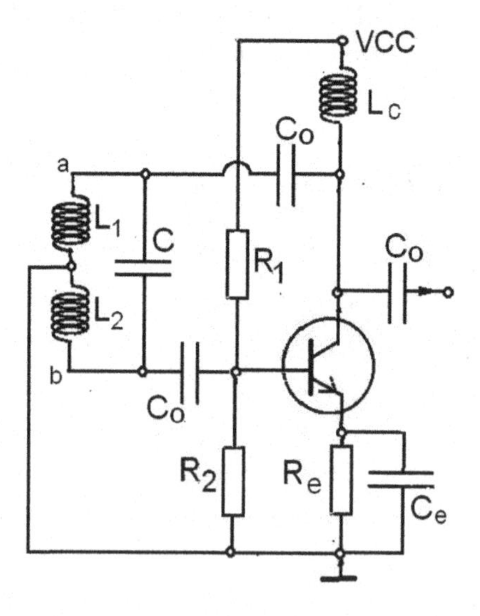


**Fig. 4.3** Hartley oscillator

### 4.1.4 Cristal Oscillators

The crystal capacitors are based on the piezoelectric effect in a piece of cristal, where mechanical forces produce electrical charges and vice versa. The piece is cut from a piezoelectric material like quartz. The natural resonant frequency of the crystal works like a resonant in the equivalent LC circuit. So, in the oscillator applications, the crystal element is used instead of the corresponding LC circuit. The crystal capacitor has a highly stable resonant frequency.

Figure 4.4 shows the symbol of a crystal capacitor and the corresponding equivalent circuit. Here, $C$ is the vibration capacitance and $C_p$ capacitance of the holder, $C_P >> C$. The capacitor $C$ determines the resonant frequency $f = \frac{1}{2\pi\sqrt{LC}}$. $R$ is a small internal resistance.

Figure 4.5 depicts a Colpitts oscillator controlled by a cristal capacitor.

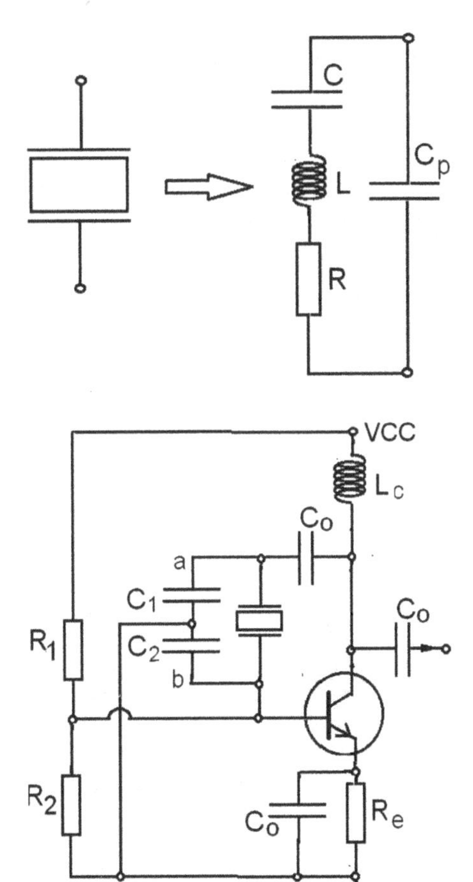


**Fig. 4.4** A cristal capacitor

**Fig. 4.5** Colpitts oscillator with cristal capacitor

## 4.2 RC Oscillators

LC oscillators for low frequencies may result little practical because of the size and cost of big inductances. A common type of oscillators for acoustic and lower frequencies are RC oscillating circuits.

The general structure of the oscillator is an amplifier with positive feedback that works for a desired frequency. If we want to generate a sinusoidal signal, the feedback element with gain H, must provide a signal with opposite phase to the input signal (inverted sign). The amplifier must have the gain equal to (or slightly greater) than 1/H. Below we will discuss some examples of such RC oscillators.

### 4.2.1 Phase-Shift Oscillator

First, take a look at the circuit of Fig. 4.6. The circuit consists of three low-pass RC circuits connected in series. The transfer function of such circuit is not just the product of there single RC circuits because of the backward interaction of the consecutive stages.

Figure 4.7 shows the results of the circuit simulation. Input signal, applied at node $v_0$ is marked with 1, and the consecutive node voltages as 2,3, and 4, respectively. It can be seen that the output voltage $v_4$ has a negative phase shift with respect to $v_1$ equal to, approximately $\pi$ radians, 180°. Using this circuit in feedback, we obtain the positive feedback.The decrease of amplitude between $v_1$ and $v_4$ must be compensated by the amplifier gain.

Circuit parameters are as follows: $R = 10\,\text{k}\Omega$, $C = 0.1\,\mu\text{F}$, $f = 100\,\text{Hz}$, $\omega = 628.32$, final simulation time equal to 0.03 s (Fig. 4.7).

Figure 4.8 depicts the voltages with curves normalized to [0, 1].

Figure 4.9 shows similar RC circuit that provides positive phase shift. This circuit may also produce the positive feedback, where the signal phase is equal to +180 degree. This circuit is commonly used, perhaps because it does not transmit the DC component of the signal.

In Fig. 4.10 we can see an RC oscillator with positive phase shift circuit. The transistor should be a small current amplifier transistor. The gain of this, one-stage amplifier has a sufficient amplification to make the circuit oscillate.

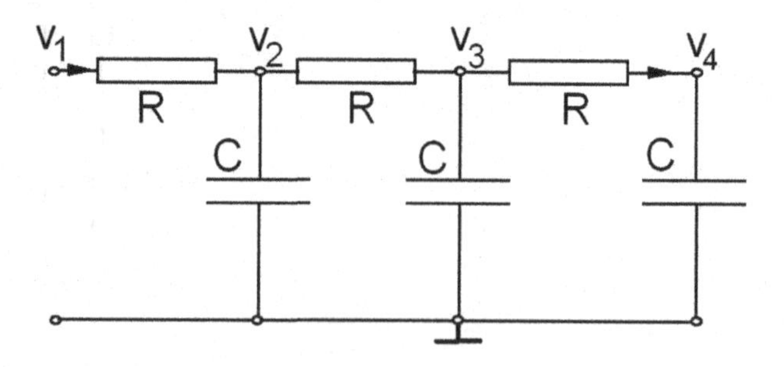


**Fig. 4.6** A phase-shift circuit

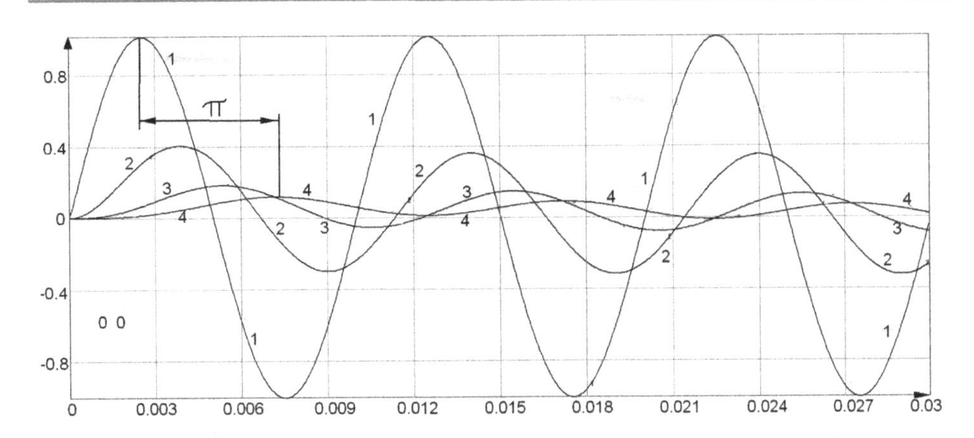


**Fig. 4.7** Phase shift of sinusoidal signal

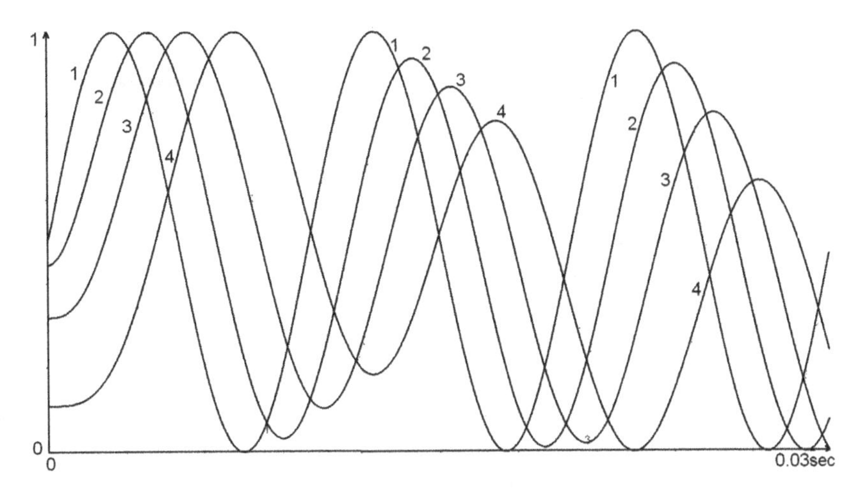


**Fig. 4.8** Phase-shift, normalized curves

**Fig. 4.9** Positive phase-shift

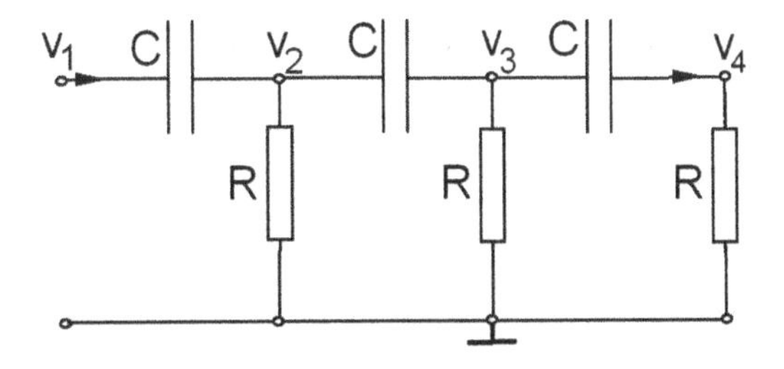

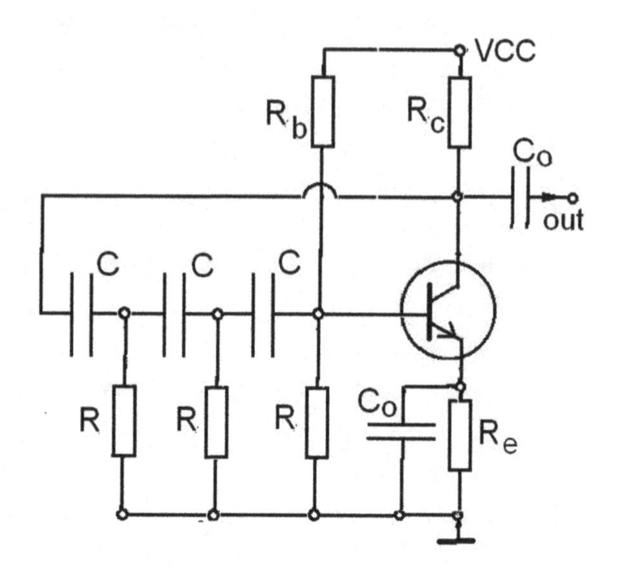


**Fig. 4.10** RC phase-shift oscillator

The signal from the transistor collector is passed to the phase-shift circuit. The capacitors C prevent the DC component to pass to the base. The operation point of the transistor is defined by the resistance $R_b$ and the resistance $R$ of the last stage of phase-shift circuit. The capacitors $C_o$ should have big capacitance.

The frequency of the signal generated by this circuit is given as:

$$f = \frac{1}{2\pi RC\sqrt{6}}$$

Figure 4.11 shows an RC phase-shift oscillator with op-amp. As the amplifier has big gain, the additional feedback resistance $R_o$ is used. It should be adjusted to achieve stable oscillations with little signal distortion.

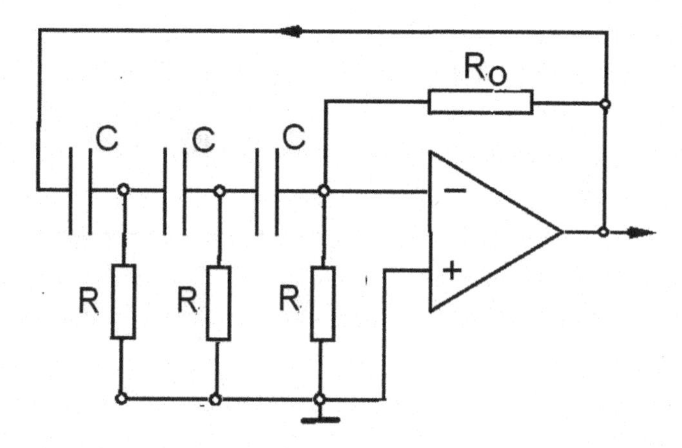


**Fig. 4.11** RC phase-shift oscillator with op-amp

### 4.2.2 Oscillator with Wien Bridge

The properties of the Wien bridge has been discussed in Sect. 3.2.4. Recall that this is an RC circuit that can be used as a second order RC band-pass filter. The frequency response of the Wien bridge are shown in Figs. 3.13 and 3.14.

As it is a bandpass filter, it can be used as the feedback circuit in RC oscillators. Figure 4.12 shows the oscillator with op-amp. It can be seen that the output signal of the amplifier is applied to the bridge formed by $R_1$, $C_1$ and $R_2C_2$. The positive feedback is taken from the common point of the two parts of the bridge. This makes the circuit oscillate. The negative feedback formed by $R_3$ and $R_4$ is used to adjust the amplifier gain, so that the oscillations remain stable with minimal distortion.

Other version, based on transistors, is shown in Fig. 4.13. Note that there is a two-stage amplifier, to provide the sufficient gain. The feedback goes through the serial part of the bridge and is applied to amplifier input from the connection of serial and parallel part.

The full Wien bridge is formed by $R_1$, $C_1$, $R_2$, $C_2$, $R_3$ and $R_4$. The difference of voltages between points a and b of the bridge is applied between emitter and base of transistor $T_1$.

The frequency generated by this oscillator is

$$f = \frac{1}{2\pi\sqrt{R_1 R_2.C_1, C_2}}$$

The Wien bridge oscillator has a gran advantage, compared to other oscillators. Note that if we change simultaneously $R_1$ and $R_2$, preserving the ration $R_1/R_2$ then the gain for the band frequency does not change, but the frequency changes. So, using a high precision double variable resistor (a part of the potenciometer), we can change frequency continuously, without changing other parameters of the circuit.

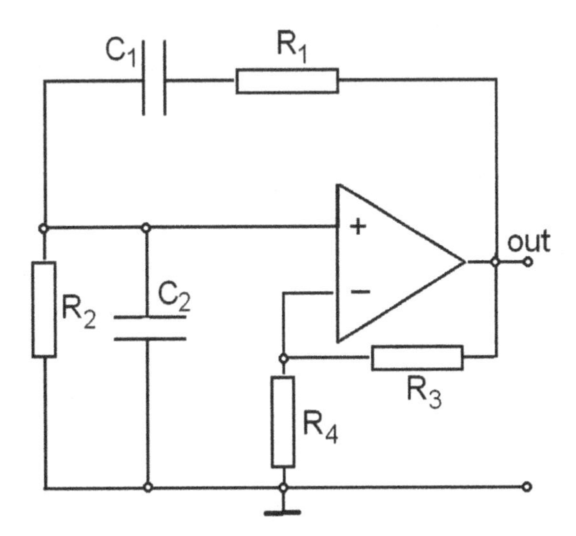


**Fig. 4.12** Oscillator with Wien bridge and op-amp

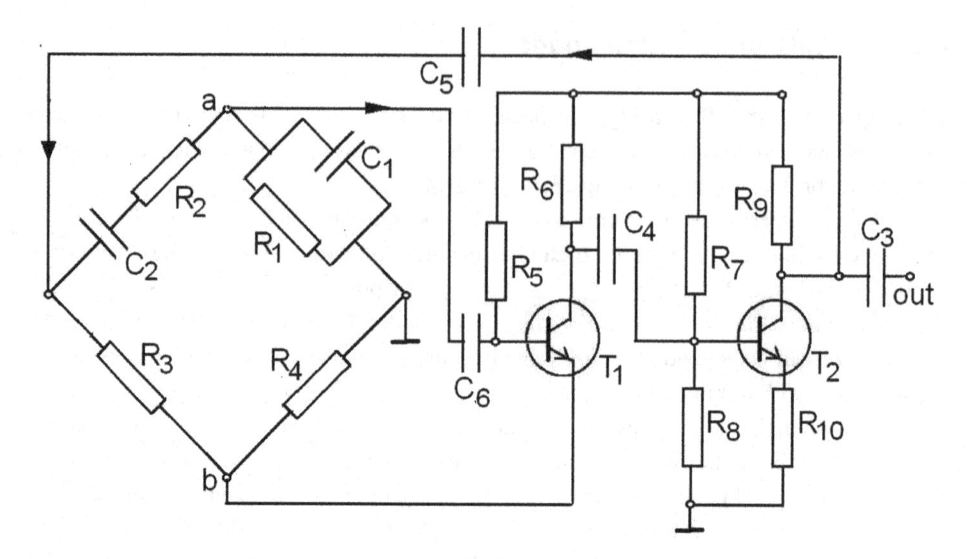


**Fig. 4.13** Oscillator with Wien bridge with transistors

### 4.2.3 Twin T Oscillator

In Fig. 4.14 there is the scheme is a twin T oscillator with op-amp. Recall that the twin T is a band-suppressing filter (see Sect. 3.2.5, Figs. 3.17 and 3.18). What is used in the oscillator is the phase shift of the circuit, rather than its gain.

In the scheme of Fig. 4.14, the twin T circuit is used as negative feedback. The circuit has also a positive feedback firmed by $R_1$ and $R_2$ that forces the start-up and maintains the oscillations.

The generated frequency for this circuit is

$$f = \frac{1}{2\pi RC}$$

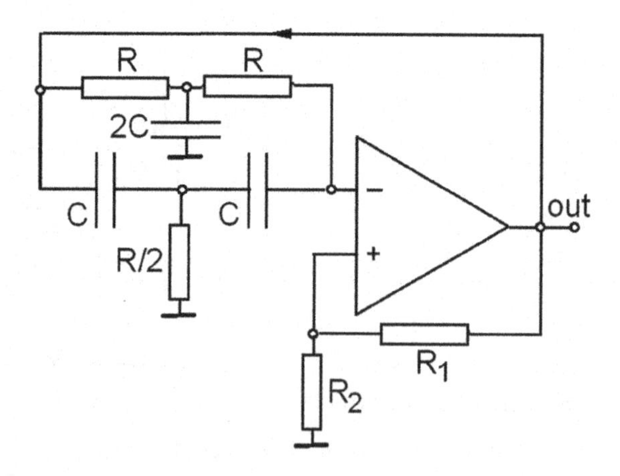


**Fig. 4.14** Twin T oscillator with op-amp

## 4.3 Multivibrators

Consider the circuit of Fig. 4.15.

Below, we assume $R_1 = R_4$, $R_2 = R_3$ and $C_1 = C_2$.

Suppose that the transistor $T_1$ conducts and the other does not. So, the voltage $v_1$ is low, and $v_2$ is high (no current $i_2$), $T_2$ is closed, so the voltage $b_2$ is low. By "low" we mean voltage that approximates zero, and "high" is a voltage near VCC. This means that the capacitor $C_1$ is charged, and its voltage is $v_1 - b_2$, both voltages low. The point $b_2$ is connected to VCC through $R_2$, so the current of $R_2$ is charging $C_1$ ($b_2$ grows). When $b_2$ reaches a level sufficient to open $T_2$,then $T_2$ starts to conduct, and voltage $V_2$ decreases. This decrement passes through the capacitor $C_2$ to base of $T_1$, and $T_1$ becomes closed. Now, $T_2$ conducts and $T_1$ closes. The voltage of capacitor $C_2$ starts increasing because of the current of $R_3$. When $C_2$ receives sufficient charge, the $T_1$ opens. This cycle repeats, and the circuit vibrates. The swap between transistors is very fast, but the process of charging the capacitors is not. This is the exponential curve, being the response of the RC circuit.

As the result, we get a rectangular periodic signal that can be taken from points $v_1$ or $v_2$. The analysis of the transient process of charging capacitor gives the formula to the frequency of the generated signal, as follows.

$$f = \frac{1}{2RCln(2)} \tag{4.2}$$

where $R$ is the charging resistor ($R_2$ or $R_1$) and C is the capacitance $C_1$ or $C_2$.

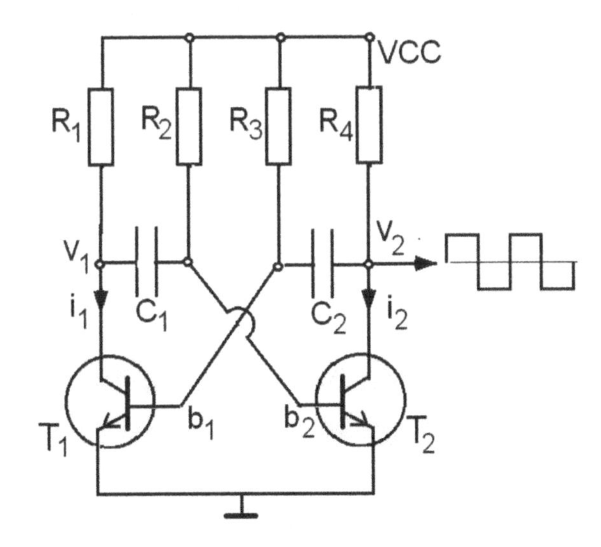


**Fig. 4.15** A multivibrator

### 4.3.1 Relaxation Multivibrator

The simplest and mostly used version of this multivibrator is based on the UJT transistor (see Sect. 1.10.4). In Fig. 4.16 we can see the scheme.

Suppose that we start with voltage of capacitance $C$ equal to zero. So, there is no current $i_e$. Then, the current grows slowly. When the emitter voltage reaches the rupture level $V_p$ (Fig. 1.26), then the emitter-B1 resistance becomes negative, and the current grows, discharging the capacitor. Then the process repeats.

As the result, we obtain a series of approximately triangular pulses at the capacitor $C$:

In Fig. 4.17 there is an example of a relaxation oscillator with op-amp.

Recall that the op-amp has very big gain K, so it can be also used as comparator. If $V_1 > V_2$ then the output voltage $V_0 = -VCC$, otherwise it is equal to $VCC$.

Let us start with $V_1 = 0$, $V_0 = VCC$ (no charge at the capacitance). We have $V_2 = V_0 R_2/(R_1 + R_2)$. The difference $V_2 - V_1$ is positive, so the op-amp remains in saturation state $V_0 = VCC$. The current $i_c$ is positive, starting with $i_c = V_0/R_3$. The capacitor is charging, and the voltage $V_1$ grows. When $V_1$ becomes greater than $V_2$, the output voltage $V_0$ switches to $-VCC$ and $i_c$ changes direction. Capacitor voltage $V_1$ starts to decrease. When it becomes lower than $V2$, the state of op-amp output switches to $+VCC$ . Current $i_c$ becomes positive, and the capacitance voltage $V_1$ grows, as in the initial state, After this, the cycle repeats. What we receive at $V_0$ is a sequence of rectangular pulses. It is easy to check that is we start with $V_1 = 0$, $V_0 = -VCC$, the circuit also enters in oscillations. Note that we cannot start with $V_1 = 0$, $V_0 = 0$, $V_2 = 0$. If so, the output would be theoretically zero, but this state is unstable because of the positive feedback through $R_1$. With any infinitesimal disturbance, the op-amp must switch to $V_0 = VCC$ or $V_0 = -VCC$.

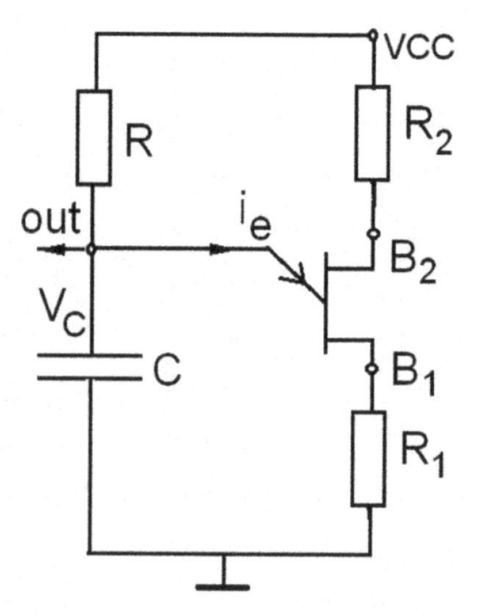


**Fig. 4.16** Relaxation oscillator with UJT transistor

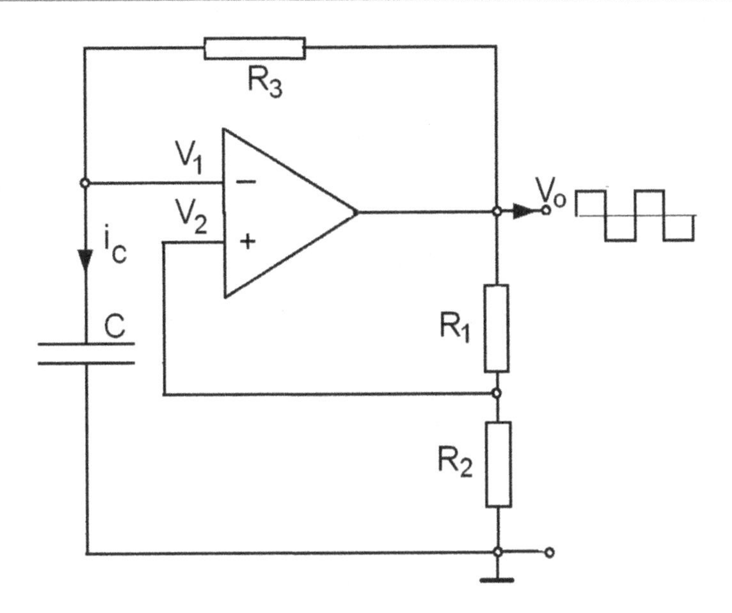


**Fig. 4.17** Relaxation oscillator with op-amp

### 4.3.2 Schmitt Trigger

Figure 4.18 shows the Schmitt trigger that switches between low (near zero) and high (VCC) output voltage depending on the input voltage $V_{in}$. The trigger has certain hysteresis, so the trigged voltage remains some time until there is a sufficient change in the input signal.

Suppose that $V_{in}$ is low. This means that the transistor $T_1$ is closed and the voltage $V_1$ is high. The base of $T_2$ receives a current that opens the transistor. The current $i_2$ grows that increases the voltage $V_2$ and maintains the transistor $T_1$ closed. Now, let us increase $V_{in}$. When this voltage becomes greater than $V_2$, the current $i_1$ of $T_1$ increases and $V_1$ decreases. The base of $T_2$ receives less current that makes the voltage $V_2$ decrease. This opens even more the transistor $T_1$ and $T_2$ closes. This provides high voltage at $V_{out}$.

Now, if we slightly decrease $V_{in}$, the $T_1$ remains open because of the increase in $V_2$. The $T_1$ will close if the change of $V_{in}$ is sufficiently big. This makes the trigger switch to high $V_1$ with different (lower) value of $V_{in}$. As the result, the circuit switches with certain hysteresis, see Fig. 4.19a. Part B of the figure shows the symbol of a trigger with hysteresis.

In Fig. 4.20 we can see the scheme of a similar trigger with hysteresis, based on an operational amplifier.

Triggers with hysteresis are used to form signal of rectangular shape. They do not trigger with small changes of input signal, so may be used to eliminate undesired perturbations and form a "clean" rectangular waves.

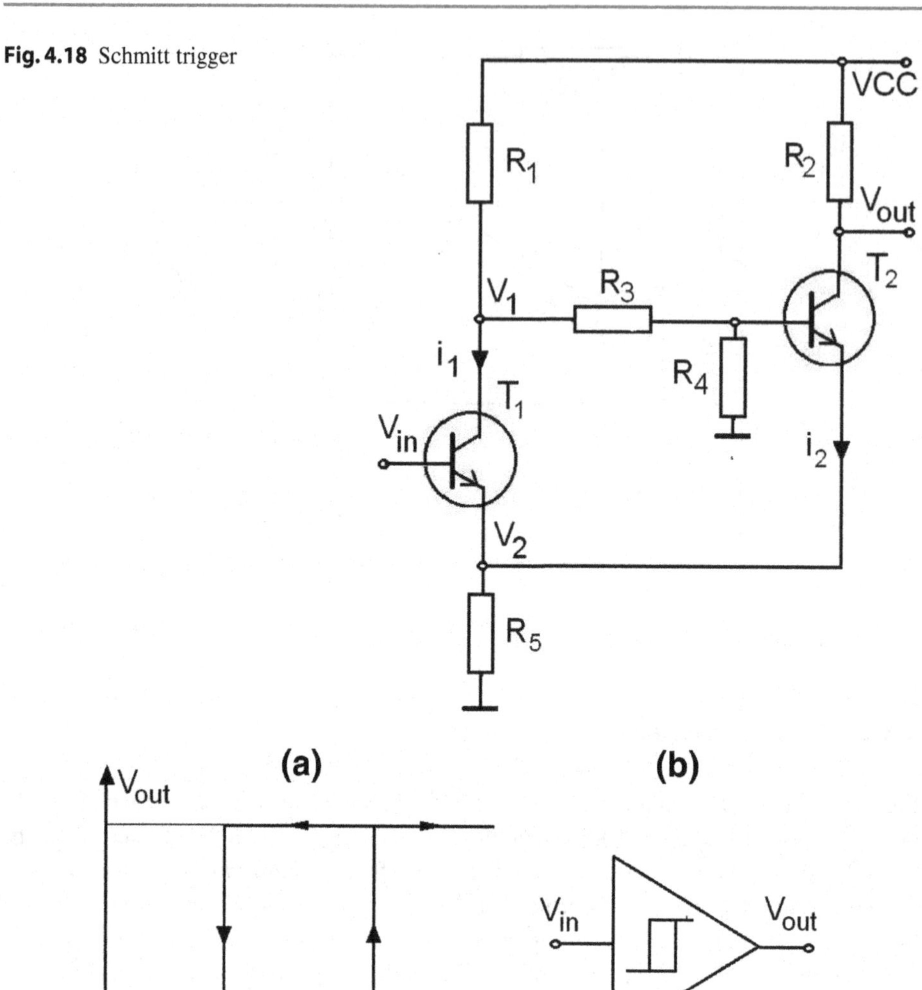


**Fig. 4.18** Schmitt trigger

**Fig. 4.19** Schmitt trigger: hysteresis

### 4.3.3 The Flip-Flop

This circuit has two stable states. It can be forced to change from one state to another. Thus, the circuit can "remember" the state, in other words, it can store one bit of information, "0" or "1".

Look at a flip-flop of Fig. 4.21. Suppose that transistor $T_1$ conducts. The current $i_1$ is big, and the voltage $V_1$ is small. This means that the other transistor receives little base

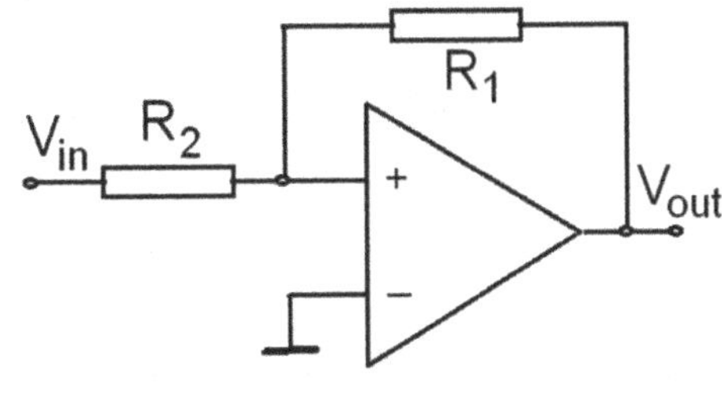


**Fig. 4.20** A trigger with histeresis, based on op-amp

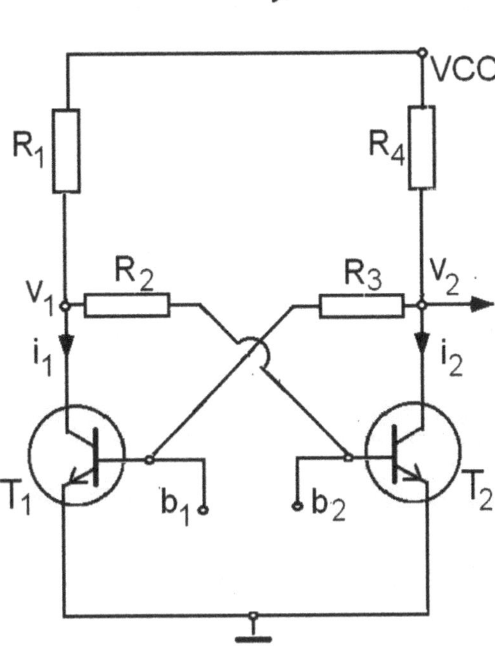


**Fig. 4.21** Flip-flop

current, and is closed. This makes the voltage $V_2$ high. This voltage is passed to $T_1$ through the resistor $R_3$, and maintains the $T_1$ open with high collector current. So, this state (S1) is stable and may remain for a long time. As the circuit is symmetric, if we suppose that $T_2$ conducts, $T_1$ must be closed, and we get another stable state (S2).

An implementation of a flip-flop with logic gates is discussed in Sect. 8.2.3.

Now, consider the state (S1). If we apply a negative pulse at $b_1$, then $T_1$ closes, and the circuit switches to state (S2). This way, the flip-flop remembers where the last negative pulse has been applied.

# Power Sources, Voltage Regulators, Thyristor Applications

# 5

## 5.1 Rectifiers

If we use the electric power provided by the AC supply red, we must convert AC to the DC voltage or current that can be used by our circuit. Figure 5.2 shows a simple rectifier where one diode and one capacitor are used. The capacitor added in part B of the figure accumulates charge and maintains it during approximately half period. Note that if the input AC (effective) voltage is equal to $V$, then the maximal rectified DC voltage is $V\sqrt{2}$.

### 5.1.1 Half- and Full-Wave

The rectified voltage decreases when the diode does not conduct. This voltage drop (ripple voltage) is equal to (see Fig. 5.1):

$$\Delta V = \frac{V_{Cmax}}{f R_L C}$$

where it is assumed that $RC >> T$. where $f = 1/T$ is the wave frequency [Hz]. Figures 5.2 and 5.3 show the ripple voltage and the diode current, respectively.

Figure 5.4 shows a *full-wave rectifier*. A transformer Tr is used with divided secondary winding. There are two diodes that produce positive DC voltage at the capacitor C- The voltages $V_1$ and $V_2$ have opposite phase, so each diode works is different half of period. The output voltage undulates with frequency $2f$ and the ripple voltage is, approximately,

$$\Delta V = \frac{V_{Cmax}}{2 f R_L C}$$

© The Author(s), under exclusive license to Springer Nature Switzerland AG 2023

S. Raczynski, *How Circuits Work*, Synthesis Lectures on Engineering, Science, and Technology, https://doi.org/10.1007/978-3-031-34934-8_5

**Fig. 5.1** Half-wave rectifier

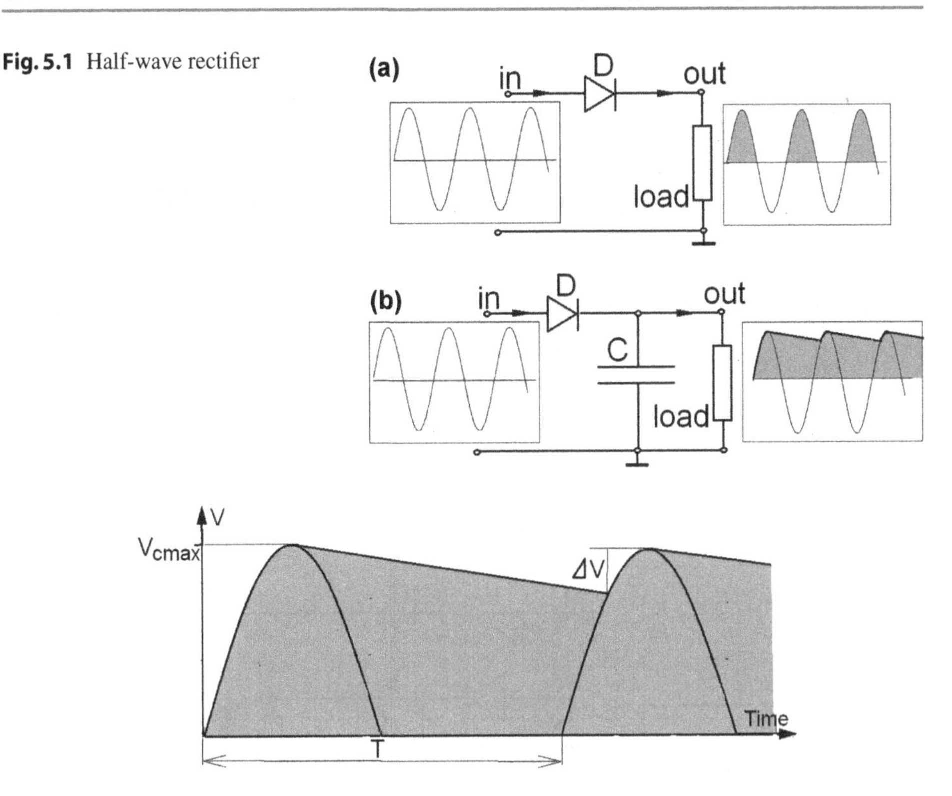


**Fig. 5.2** Half-wave rectifier, ripple voltage

The average diode current is equal to the load current. However, note that the root mean square (rms) value that produces energy loss on the diode is much more greater than the average current.

Now, look at Fig. 5.3. The diode current is not sinusoidal. In the figure, diode current $i(t)$ is approximated by a rectangle of width $cT$, and height $i_m = I_{av}/(c)$, where c is a coefficient $c << 1$. This makes the average current equal to $1/T \times cT \times (I_{av}/c) = Iav$. So the rms value is

$$I_{rms} = \sqrt{\frac{1}{T}\int_0^T i^2(t)dt} = \sqrt{(1/T)Tc\,(I_{av}/c)^2} = \frac{I_{av}}{\sqrt{c}} \tag{5.1}$$

For example, if $c = 0.1$, then the rms current is equal to 3.162 $I_{av}$. If the diode current were equal to $I_{av}$ (DC) in the whole period T, then the rms would be equal to $I_{av}$. This means that the diode with nominal maximal average current may not support the current of our rectifier.

Another full-wave rectifier can be seen in Fig. 5.5. This rectifier may work with or without transformer. If the input voltage is in the first half-period ($V_a > V_b$), then the diodes $D_1$ and $D_2$ conduct. Otherwise. the diodes $D_3$, $D_4$ open. So, we have the full-wave rectification.

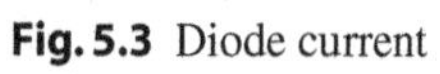

**Fig. 5.3** Diode current

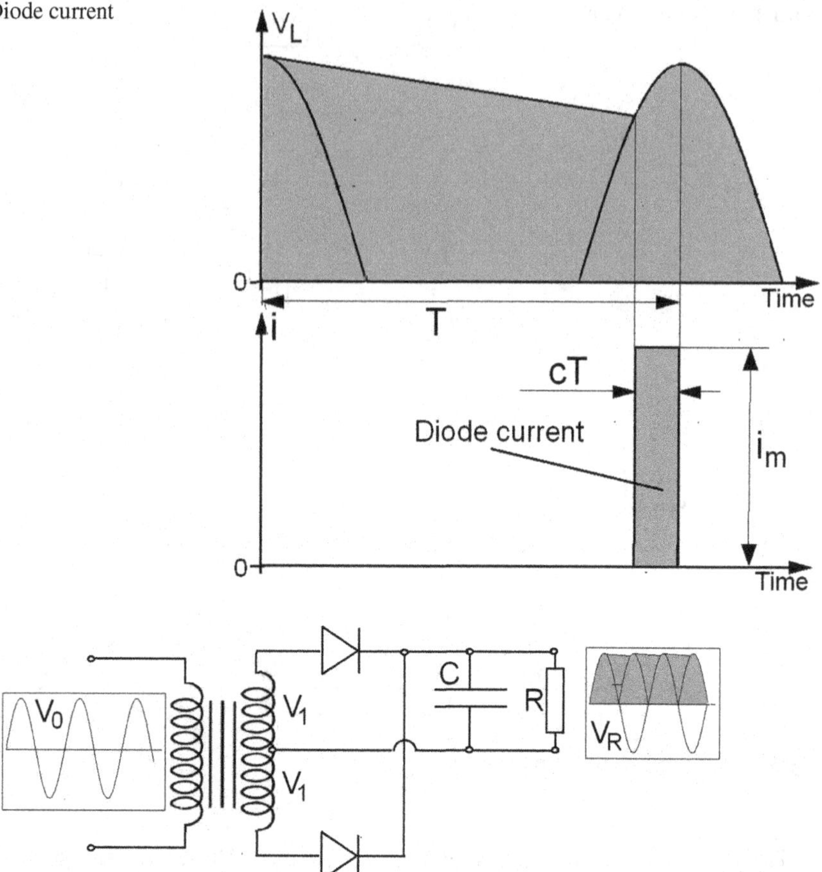


**Fig. 5.4** Full wave rectifier

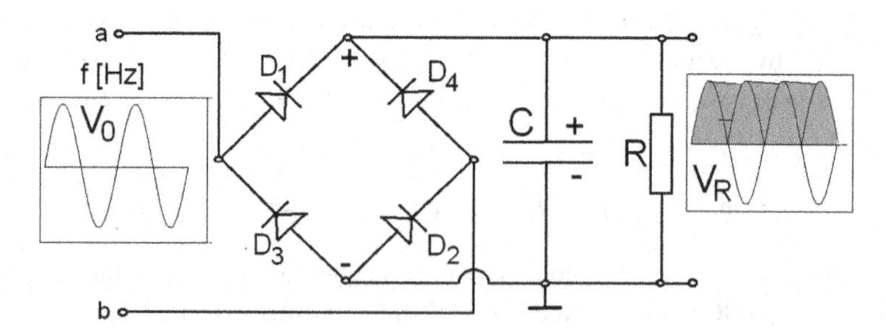


**Fig. 5.5** Full wave bridge rectifier

Let us make a comment on the supply sources. Suppose that we use an ideal transformer and ideal capacitance. Now, suppose that we connect the supply circuit to the (ideal) AC voltage source, at a time instant when the supply voltage is different from zero. This means that we connect an ideal voltage source to an ideal capacitor. Recall that the capacitor voltage is

$$V(t) = \frac{1}{C}\int_0^t i(t)dt \text{ this means that } i(t) = C\,dV/dt$$

where $i(t)$ is the capacitor current. Note that we turn on the circuit to a non-zero voltage, which means that $di/dt/$ is infinite. If the devices we use are real (source with internal resistance, etc.), this initial pulse of current is finite, but it may be quite big. The disconnecting of the circuit may produce even worse results because the interruption of current flowing through an inductance produces an infinite voltage pulse. To avoid such problems, we can use a "zero cross-over" circuits that will be discussed in Sect. 5.2.4.

### 5.1.2 Voltage Multiplier

If we use a transformer at the input of the rectifier, then we can design it to get a required DC voltage. If the source has a predetermined voltage and the required voltage is higher, then *voltage multiplier circuits* can be used. The simplest circuit of this kind is shown in Fig. 5.6.

If the input voltage is $V_0$ (effective) sinusoidal, then it is easy to see that the voltage $V_1$ must oscillate between zero and $V_1 = 2V_0\sqrt{2}$. So, the diode $D_2$ charges the capacitor $C_2$ to DC voltage $V_2 = 2V_0\sqrt{2}$.

Using more diodes and capacitors we can multiply the input voltage several times.

### 5.1.3 DC Converter

Figure 5.7 shows an example of *boost DC converter* that produces high voltage, receiving a DC supply. The idea of such circuits is to take advantage of the fact that interruption of

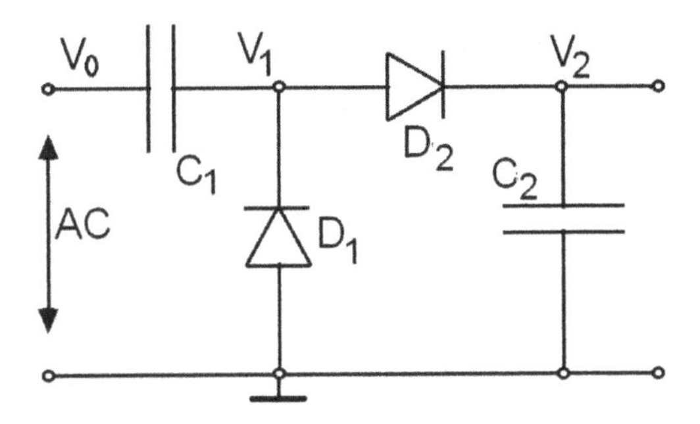


**Fig. 5.6** A simple voltage multiplier

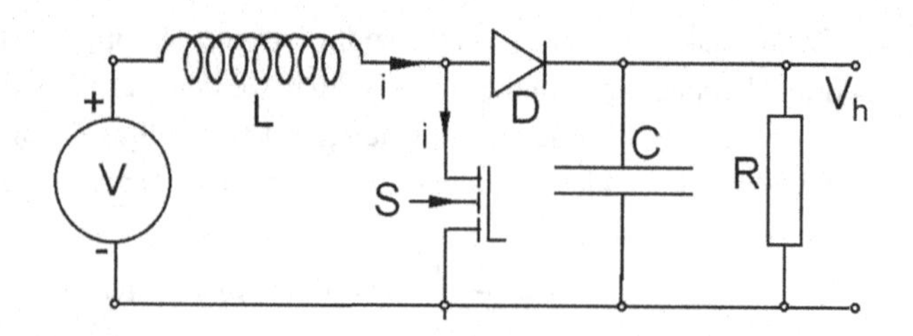

**Fig. 5.7** Boost DC converter

a current flowing through an inductor produces big voltage pulse. In Fig. 5.7 the element S is a power MOSFET that works as a switch. In fact we can use any other device that can switch (interrupt) current. It works as follows.

First, The switch S is closed. After connecting the DC (low voltage) source V, the voltage on capacitance C grows, and reaches the level V. Then, the current i flows through the inductor L and the closed switch S. When the switch opens, the current is interrupted, and on L appears a big voltage pulse. This pulse passes through the diode and increase the charge of the capacitor C. The control voltage applied to the MOSFET must be a rectangular wave that makes the switch periodically close and open. The switching frequency must be greater than the cutoff frequency of the circuit RC (Fig. 5.7). In commercial converters, this frequency is normally greater than 400 kHz.

Additional voltage regulators can stabilize the output voltage. This voltage can be controlled by the duty ratio of the controlling rectangular wave and by other circuit parameters.

## 5.2 Voltage Regulators

### 5.2.1 Zener Diode Regulator

The simplest voltage regulator with a Zener diode is shown in Fig. 5.8.

The output voltage $V_2 = V_z$ is the stable voltage of the Zener diode Z, current $i_z$ flows through the diode, and $i_2$ is the load current.

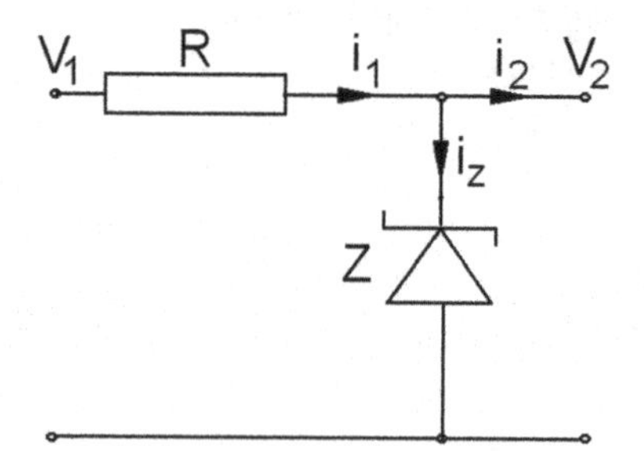

**Fig. 5.8** Simple voltage regulator

This circuit works under the following conditions:

* $V_1 > V_2$
* $R < (V_1 - V_z)/i_2$
* $V_1/R < i_{zmax}$

where $i_{zmax}$ is the maximal diode current.

This circuit can be used as a voltage reference, but it is hardly useful as a supply source, unless the load current is very small.

### 5.2.2 Regulator with Transistors

In Fig. 5.9 there is a scheme of a voltage regulator with a power transistor $T_1$. It works as follows.

First, suppose that the circuit is in a steady state. $V_1$ is the supply voltage (e.g. provided by the rectifier), and $V_2$ is the voltage on the load. Also suppose that the load impedance is real and the circuit is stable. Now, if $V_2$ decreases for any reason (a change in load resistance), then the voltage divisor $R_2$, $R_3$, $R_4$ makes the voltage of the base of $T_2$ decrease. This means that $i_e$ decreases (the voltage $V_z$ is fixed). The voltage drop on $R_1$ decreases also that makes the current of $T_1$ grow, returning $V_2$ to desired value.

The desired voltage $V_2$ can be changed by changing the position of the potenciometer $R_3$. Supposing that the base-emitter voltage drop of $T_2$ is equal to 0.6 V and that the $\beta$ of $T_2$ is big, the regulated voltage $V_2$ is as follows.

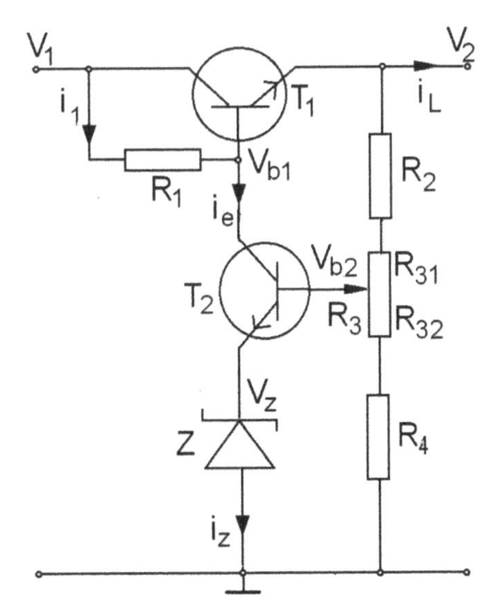


**Fig. 5.9** Voltage regulator with transistors

$$V_2 = (V_z + 0.6)\frac{R_2 + R_3 + R_4}{R_{32} + R_4}. \tag{5.2}$$

In this circuit, transistor $T_2$ should be a signal amplifier transistor with big $\beta$, and $T_1$ should be a power transistor that supports the maximal load current $i_{Lm}$, and $T_2$ must support the current $i_{Lm}/\beta_1$ ($\beta$ of $T_1$).

### 5.2.3 Regulator with Op-amp

A regulator with op-amp is shown in Fig. 5.10. The amplifier works as the transistor $T_2$ of Fig. 5.9. It can be seen that there is a negative feedback from the load voltage. The value of voltage $V_2$ is the same as in circuit Fig. 5.9. The difference is that the op-amp has great gain, and the voltage control is more exact. However, remember that the op-amp need the supply voltage $\pm$ VCC to operate.

There exists a series of integrated circuits that implement similar voltage regulator method, like LM78xx of Fairchild (xx is the voltage: : 5, 6, 8, 9, 10, 12, 15, 18 or 24 V). These devices are easy to use, they need no additional external elements to connect, see Fig. 5.11.

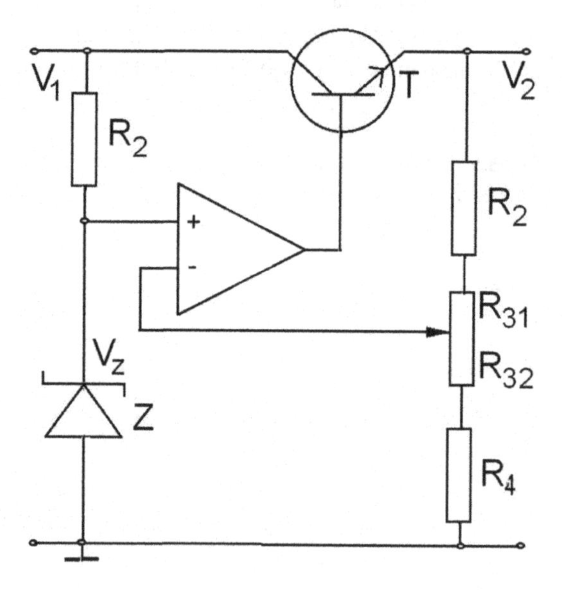


**Fig. 5.10** Voltage regulator with op-amp

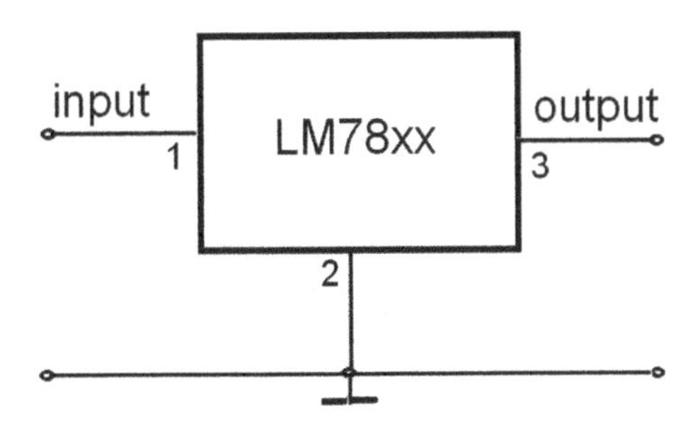


**Fig. 5.11** Voltage regulator IC

### 5.2.4 Zero-Crossing Switch

Figure 5.12 depicts the idea of zero-crossing switch. Such devices are used to avoid big current pulses when we turn on or off an electric devise. These pulses may go back to the AC power supply and affect other devices, or produce undesired noise in the speakers.

The idea of the circuit Fig. 5.11 is quite simple. There is a rectifier bridge without output capacitance. This produces the rectified wave at point c that reaches zero Volts when the supply AC voltage crosses level zero. The base of transistor T receives positive current when voltage at c is positive. When this voltage approaches zero, then the transistor closes. This produces a short positive pulse at the collector. So, we obtain a sequence of pulses that occur when the supply voltage crosses zero. The switch S can pass these pulses to a triac or other switching device that connects another device. If we close the switch any time, the triac opens with the next, nearest pulse.

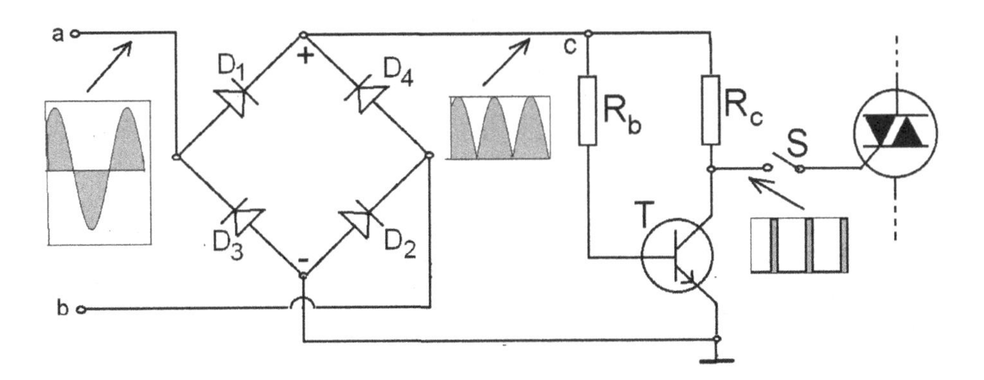


**Fig. 5.12** Zero crossing switch

# 6 Control Circuits, PID, Motion Control

## 6.1 Feedback Control

### 6.1.1 The Transfer Function

Before discussing circuits, let us repeat here the definition of the *transfer function*, mentioned in Sect. 3.1.3. We use the Laplace transform to define the transfer function, without entering the theory of the transform. Here, the important fact is that the Laplace complex variable “s” is treated as the *differentiation operator*. So, to the derivative $dx(t)/dt$ it corresponds $s\,x(s)$ in the Laplace transform. The integral $\int_0^T x(t)dt$ converts in $x(s)/s$.

Consider the differential equation with initial condition x(0)=0, $u(t)$ - given function of time.

$$a\frac{d^2x}{dt^2} + b\frac{dx}{dt} + x(t) = u(t) + d\frac{du}{dt} \tag{6.1}$$

Applying the L-transform to both sides of (6.1) we obtain

$$as^2x(s) + bsx(s) + x(s) = u(s) + dsu(s)$$

This means that the solution in terms of L-transform is

$$x(s) = u(s)\frac{1 + ds}{as^2 + bs + 1}$$

Now, define

$$G(s) = \frac{x(s)}{u(s)} = \frac{1 + ds}{as^2 + bs + 1} \tag{6.2}$$

$G(s)$ of (6.2) describes the dynamics of a device with the input signal $u(t)$ and output $x(t)$. The above function $G(s)$ is the *transfer function* of the device.

© The Author(s), under exclusive license to Springer Nature Switzerland AG 2023

S. Raczynski, *How Circuits Work*, Synthesis Lectures on Engineering, Science, and Technology, https://doi.org/10.1007/978-3-031-34934-8_6

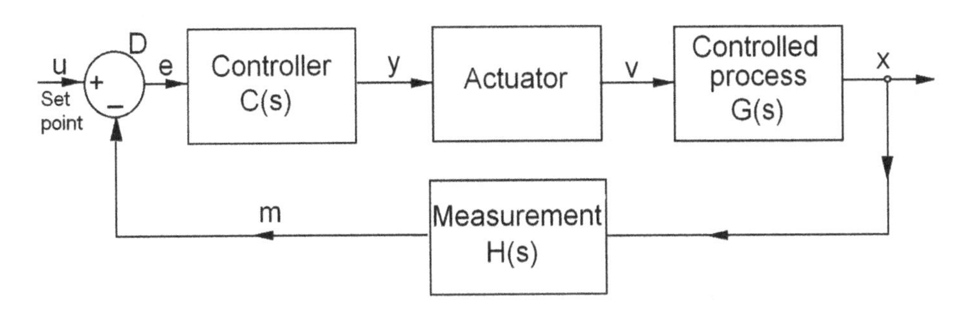


**Fig. 6.1** A feedback control system

The transfer function is the property of the corresponding device. It does not depend on input and output signals. From the control theory it is known that given a transfer function, we can get the system frequency response, simply replacing the variable $s$ by $j\omega$.

Figure 6.1 shows the general scheme of a *closed-loop (feedback)* control system.

### 6.1.2 Automatic Closed-Loop Control Circuit

The purpose of the circuit is to control a process $G(s)$ in such a way that the value of the output variable $x$ be equal or close to the *set point* value $u$.

The actual value of process output $x$ is measured by the measurement instrument with transfer function $H(s)$. The element D compares $x$ and $U$, and provides the difference $e = u - x$. It is called *control error*. The control error is processed by the *controller* $C(s)$. It produces signal $y$ that, through an *actuator* is applied to the controlled process. The actuator is frequently a non/linear element and does not have a transfer function. The controlled process can be, for example an electric heater, the measurement instrument a thermo-cupple with its amplifier, and actuator may be a power-electronic circuit with thyristors that provides the heating power to the heater.

As for the controller, there is a great variety of implementations, not only electric but also pneumatic, hydraulic or mechanical. Here, we consider electric devices only.

Supposing an ideal actuator with gain equal to 1, the overall transfer function of this circuit is as follows.

$$F(s) = \frac{x(s)}{u(s)} = \frac{C(s)G(s)}{1 + C(s)G(s)H(s)} \tag{6.3}$$

### 6.1.3 Two-Point Control

The simplest controller is just a comparator that provides "high" or "low" output, depending on the control error. It may be described by the following formula.

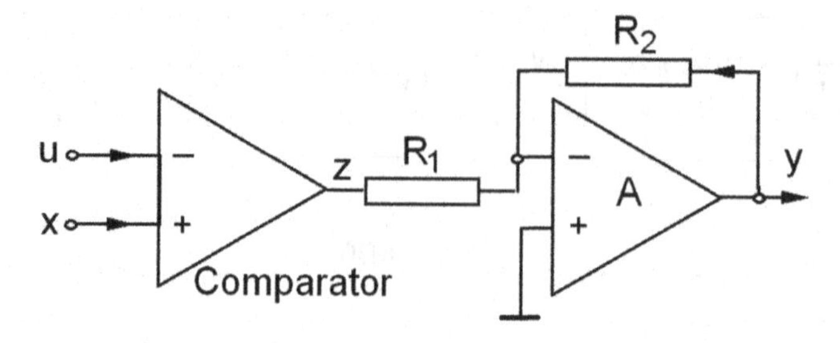


**Fig. 6.2** Two-point controller

$$y = \begin{cases} V_{max} \text{ if } e > 0 \\ 0 \text{ if } e \leq 0 \end{cases} \tag{6.4}$$

The actuator can be a thyristor-based device that connects the power to the heater.

As for the controlled process (the heater), it is a heating element that always responds with certain inertia. Suppose that the heater transfer function is a second-order inertial element with the following transfer function:

$$G(s) = \frac{P}{1 + 3s + s^2} \tag{6.5}$$

where $P$ is the process overall gain [degree/Volt]. In the following, we will not use the physical dimensions of the signals, treating them as relative values, with respect to the maximal level of the corresponding variable. So, we suppose $P = 1$ the actuator gain $A = 1$ and the controller output $V_{max} = K$, where $K$ will be called *controller gain*.

The controller can be implemented as a comparator with output amplifier, if necessary, as shown in Fig. 6.2. In this case, the controller has no transfer function. It is described by the Eq. (6.4).

Note that in Fig. 6.2 the set-point $u$ enters in the negative input of the comparator. This is because the output amplifier A also inverts the signal sign. In the simulation below we assume ideal actuator, $v = y$. The set-point $u$ is equal to 0.5, and the controller generates signal 0 or 1, in relative units (1 corresponds to $V_max$, K = 1). The overall gain of the heater is $P = 1$. The measurement instrument is supposed to respond as the first order inertial object with time-constant equal to 0.2. The corresponding transfer function is as follows.

$$H(s) = \frac{1}{1 + 0.2s} \tag{6.6}$$

The action of this controller is shown in Fig. 6.3. The plot with gray filling is the response of the controller. It can be seen that the controller switches between 0 and 1, when the measured value $m$ of $x$ crosses level 0.5 (the set-point). In these moments, the control error $e$ is equal to zero.

Two-point controller is simple and it is frequently used, when a more exact control is not needed. Observe that the process output $x$ (the controlled variable) enters in oscillations and

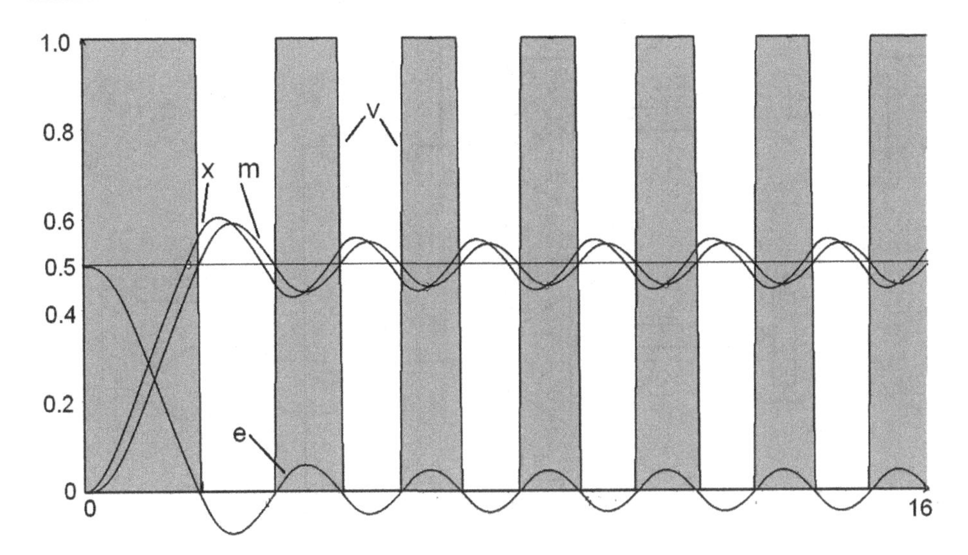


**Fig. 6.3** Two-point controller, oscillations

never reaches a steady state. The oscillations of the two-point controller may be quite big, in particular for processes of higher order or with time-delay.

### 6.1.4 PID Control

The PID controller has a continuous action and, in its basic version, it is linear and has a transfer function. Control of this kind includes three components of controller response:

* *Proportional action* (part P): The controller response is proportional to the control error $e$. If the error grows (process output $x$ decreases), then the controller response grows.

* *Integral action* I. This part of the controller is an integrator. It integrates the control error. This action is slow, but if the non-zero error exists for a long time, this action can compensate it.

* *Derivative action* D. This part of the controller reacts to the derivative of the error. If the error changes rapidly, this action generates the fast response of the controller.

A wide review of the PID theory and versions can be found in [1].

The equation of the basic PID controller is as follows.

$$y(t) = K\left(e(t) + \frac{1}{T_i}\int_0^t e(\tau)d\tau + T_d\frac{de}{dt}\right) \tag{6.7}$$

The parameter $T_i$ is called *integration or duplication time*, and $T_d$ is the *derivative time constant*.

The transfer function of the above controller is

$$C(s) = \frac{y(s)}{e(s)} = K\left(1 + \frac{1}{T_i s} + T_d s\right) \tag{6.8}$$

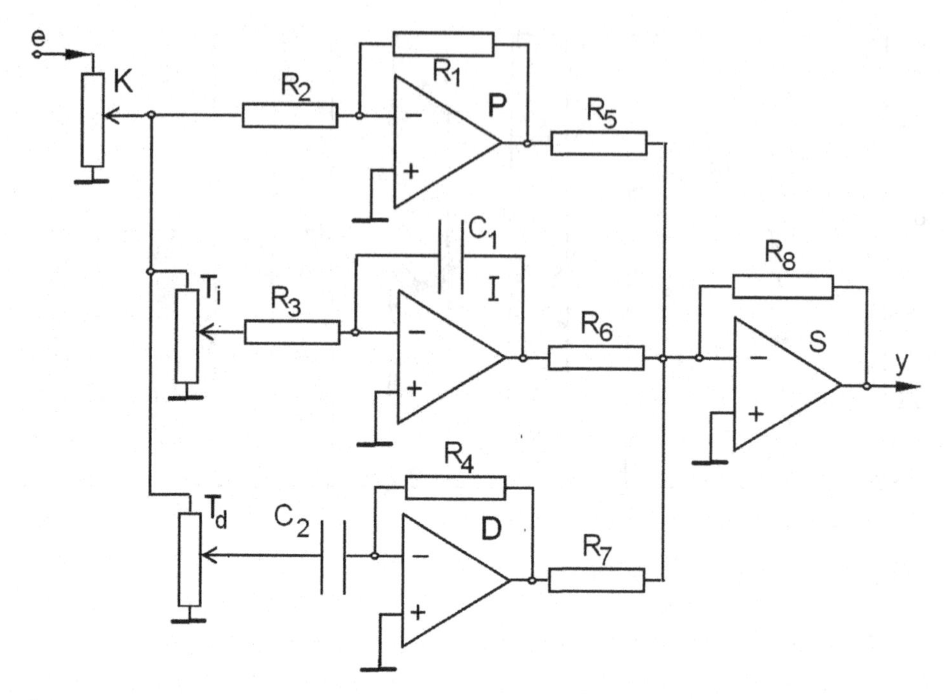


**Fig. 6.4** Basic PID controller

Figure 6.4 shows a possible implementation of the PID controller, using op-amps. The three parts of the controller are marked as P(proportional), I(integral) and D(derivative). Potenciometer K defines the overall controller gain. Potenciometers $T_3$ and $T_d$ define the share of the part I and D, respectively. The digital, sampled/data version of the PID controller will be discussed in Chap. 7.

The op-amp P works as amplifier. Op-amp I is an integrator, and op-amp D is a differentiating circuit. The op-amp S is a sumator.

The input signal to the circuit is the control error e.

Figure 6.5 depicts the results of a simulation run of the heater of Eq. (6.5), with the basic PID controller. The controller parameters are: $K = 4$, $T_i = 4$, $T_d = 3$, $H(s)$ as in Eq. (6.6), actuator gain equal to one.

Simulation of Fig. 6.5 shows plots for two settings of the controller. Curves marked as $x$ and $e$ show the response of the heater and the control error with the complete PID controller, respectively. Curves $x_p$ and $e_p$ are obtained with both integral and derivative actions disabled. The set-point is equal to 0.6.

It can be seen that if the controller is proportional only, a considerable static error remains. If we connect the integral and derivative actions, then the control quality is good. The static error disappears and the response is a little bit faster, with less oscillations.

Now, let us see a possible implementation of the actuator. In this case, this cannot be just a switch. The actuator must generate supply, preferable as AC, with power that change continuously, controlled by the voltage signal from the controller.

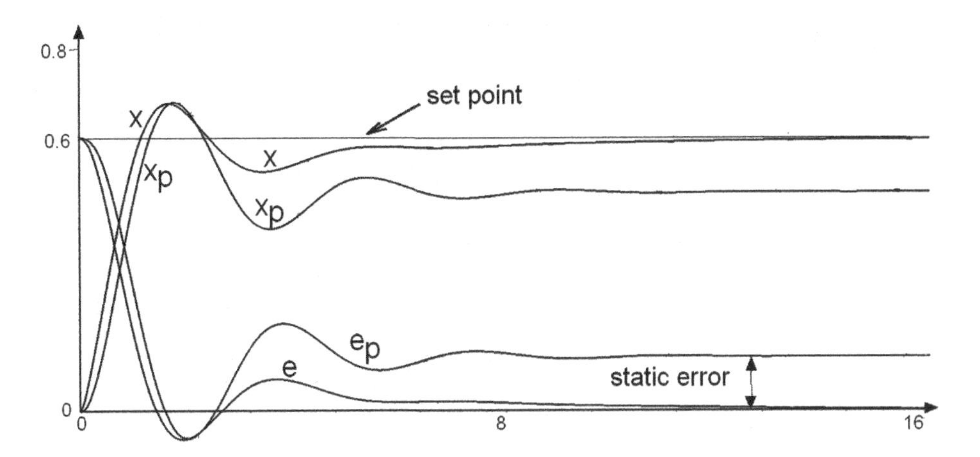


**Fig. 6.5** Basic PID controller, closed-loop control

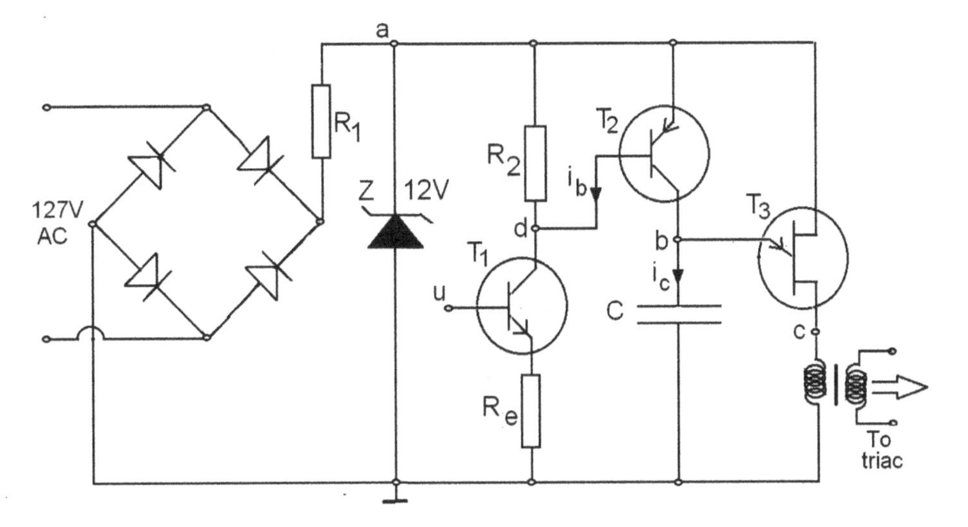


**Fig. 6.6** Power control

Figure 6.6 displays the scheme of a simple AC power control. This circuit does not produce power, but a series of pulses that can control a power supply with triacs or similar power-electronic elements. The wave/forms are shown in Fig. 6.7.

The diode bridge rectifier provides a full wave rectified voltage. Zener diode and resistor $R_1$ limit the voltage at point a to 12 V. However, this voltage returns to zero each time the supply voltage crosses zero. This makes the whole circuit shut down every half-period. When the voltage at a grows to 12V, the current of transistor $T_1$ flows, controlled by external control voltage $u$. If $u$ grows, then the collector current of $T_1$ grows, and the voltage at point D decreases. The base current of $T_2$ (PNP transistor) grows, and the capacitor C is being charged, as shown in Fig. 6.7. So, the rate of charge of C is controlled by the input signal

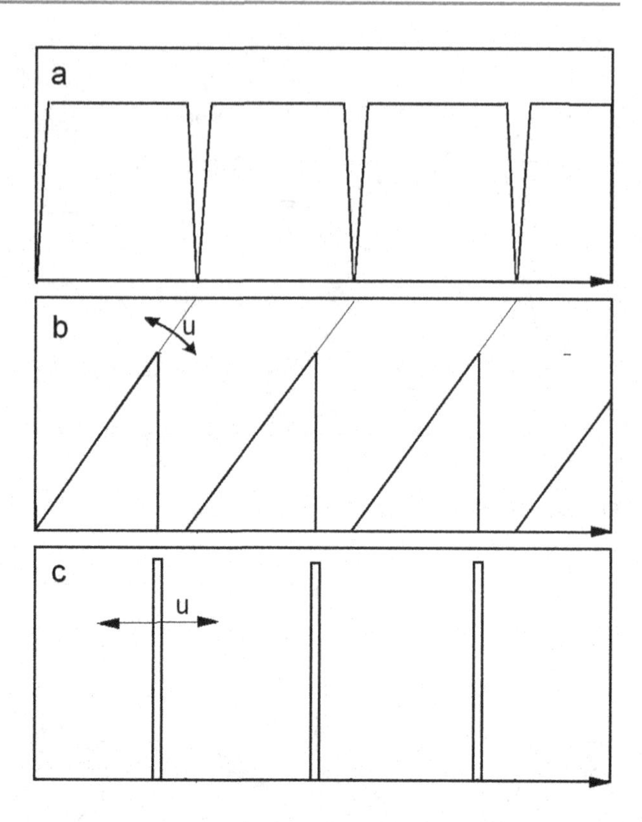


**Fig. 6.7** Power control, waveforms

u because this controls the slope of capacitor charging line. It is supposed that u changes slowly, compared to the frequency of the AC supply.

Transistor $T_3$ may be an UJT transistor or any other electronic switching device. If the voltage of the capacitor is high enough, $T_3$ switches on, and C discharges. The current pulse of $T_3$ produces the output pulse that can control a TRIAC-based AC power supply. Figure 6.8 shows the current waveform on the load. The gray part of the plot is when the current flows, and in the other part of the AC period there is no load current. This make the average load current change continuously, according to the control u.

Similar power control circuits may be used in control of electric motors, lamps, ventilators and other.

### 6.1.5 Enhanced PID

The set point may be changed any time, continuously or by step-like changes. If we rapidly change the set point, then the control error changes also. This may cause unnecessary big pulses in controller output. To avoid this, in some PID versions the set-point does not pass through the derivative part of the controller. In this case, the PID equation is as follows (Fig. 6.9).

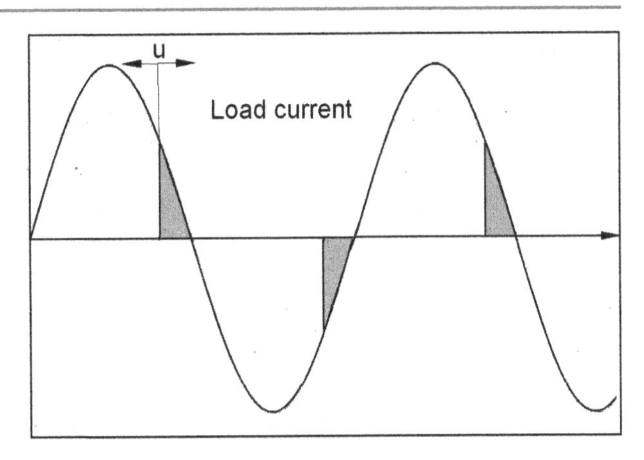


**Fig. 6.8** Power control, load waveforms

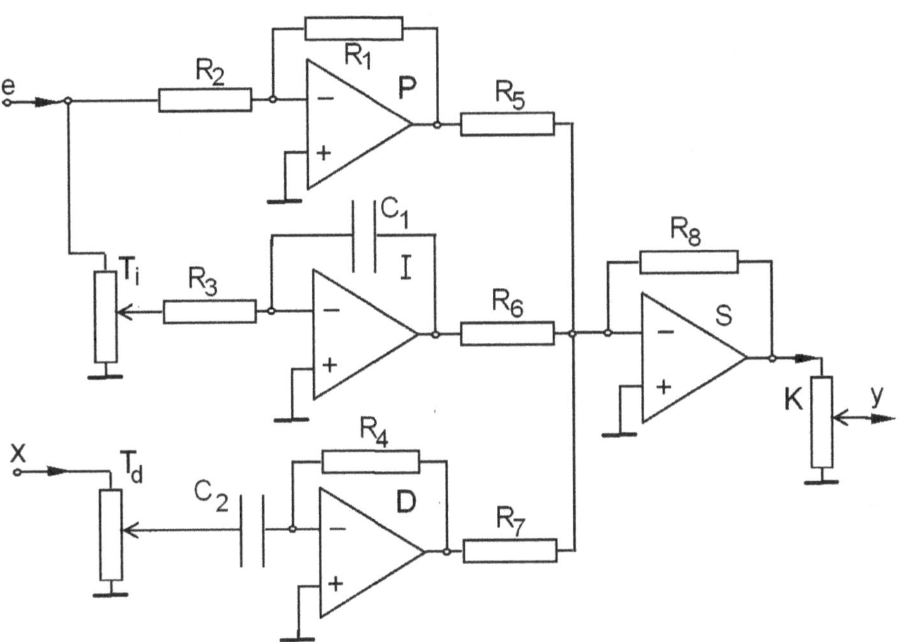


**Fig. 6.9** Modified PID

$$y(t) = K\left((u(t) - x(t)) + \frac{1}{T_i}\int_0^t (u(\tau) - x(\tau))d\tau - T_d\frac{dx}{dt}\right) \tag{6.9}$$

Another modification of the controller is a "smoothed derivative" action. The ideal differentiator is highly sensitive to high-frequency disturbances, like a measurement noise or interfering radio signals.

The modified derivative part is as shown in Fig. 6.10. A small capacitor $C_1$ is added in the feedback, and a small resistance $R_2$ is added in the input impedance. Calculating the transfer function of this circuit we obtain

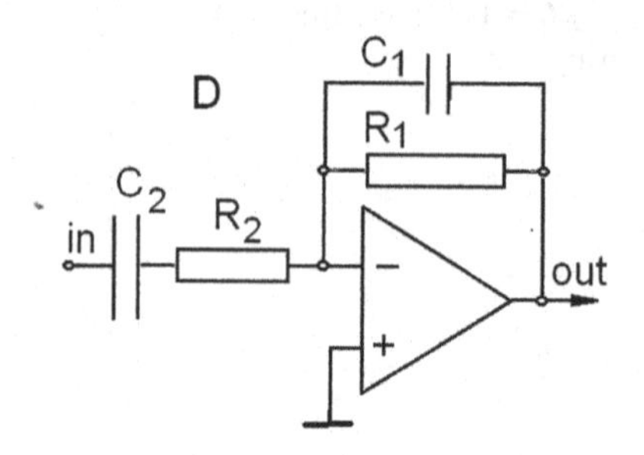


**Fig. 6.10** Modified derivative action

$$D(s) = \frac{out(s)}{in(s)} = -\frac{R_1C_2s}{R_1C_1R_2C_2s^2 + (R_1C_1 + R_2C_2)s + 1} \tag{6.10}$$

Observe that if $C_1$ and $R_2$ tend to zero, then the transfer function (6.10) reduces to $R_1C_2s$, like in the ideal differentiator.

In terms of frequency this is

$$D(j\omega) = -\frac{R_1C_2j\omega}{(R_1C_1 + R_2C_2)j\omega + 1 - R_1C_1R_2C_2\omega^2} \tag{6.11}$$

Looking at (6.11), it can be verified that for slow signal $\omega \to 0$, so $D(j\omega) \approx -R_1C_2j\omega$. However, for high frequencies, we have $D(j\omega) \approx \frac{j}{R_2C_1\omega}$ . This means that the high frequencies do not pass through the circuit.

### 6.1.6 PID with Anti-windup

One of the important problems in PID control is the effect of *windup*. This occurs when the control error is different from zero for a relatively long time interval. In this case, the integrator may accumulate a big value and saturate. After this, it takes some time to go back to the normal state, and the appropriate response of the controller is delayed.

There are various versions of PID that eliminate or decrease the windup effect. One of possible and most known analog solution is shown in Fig. 6.11. This PID includes a feedback that activates when the saturation of the actuator takes place. In this PID, A is the model of the saturation element, or the real actuator. The signal $w$ is the sum of the parts P, I and D of the controller.

First, suppose that $w$ is inside the normal operation region. Then, $y = w$, y $g = 0$. In this case, the controller response $y$ is as in a conventional PID. Observe that the signal $p$ (the set point) is not subject to differentiation.

Now, suppose that $w$ is out of the lineal operation region, and the actuator saturates. So, the signal $w$ is different from $y$ and $g$ is the integral of $y - w$. This difference is negative. It is added to $h = e/(T_is)$ and shuts down the integrator.

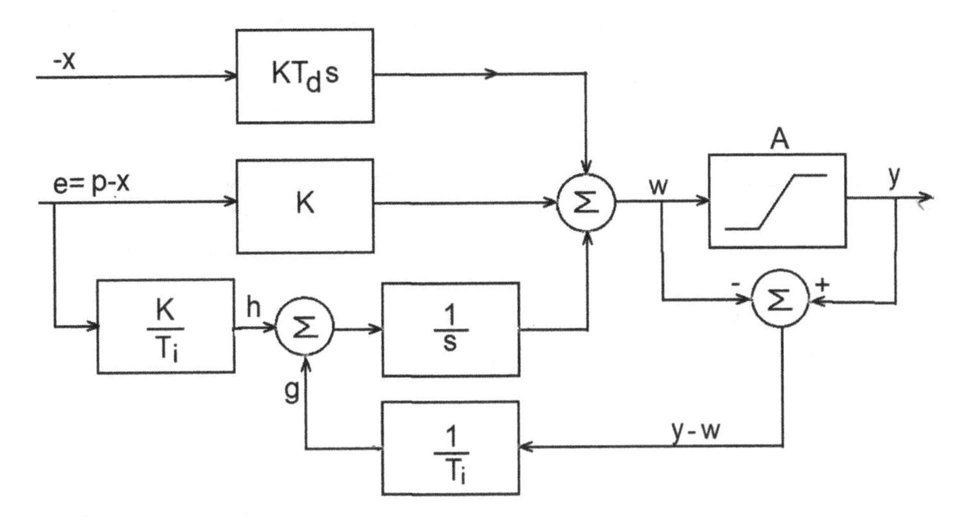


**Fig. 6.11** Anti-windup PID

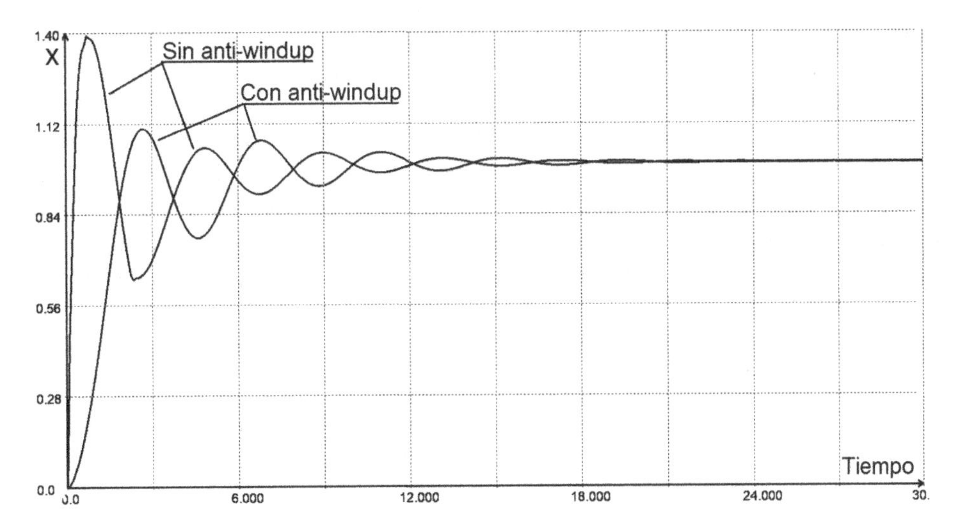


**Fig. 6.12** Anti-windup PID simulation

Figure 6.12 depicts the simulation of a control system with the anti-windup PID. The controlled process is the second order object, with the following transfer function.

The controlled process is a second-order object, with the following transfer function.

$$G(s) = \frac{1}{2s^2 + s + 1} \tag{6.12}$$

The controller gain is $K = 4$ integration time $T_i = 2$ and the derivative action coefficient $T_d = 0.2$, The set point is $p = 1$. The actuator has the saturation level $L$, and the signal that enters to the process has the saturation level $L_p$. In the anti-windup mode $L = L_p = 1.5$. To disconnect the anti-windup mechanism, we simply set $L$ to a very big value ($w$ does not reach the saturation level). Without wind-up, the value of the integrator grows rapidly. This makes the response of the controller too strong, and results in the overshoot of controlled variable approximately 40%. If we connect the windup, the process response has overshoot of 12%.

The industrial anti-windup PIDs have three input terminals: SP is the set point, MV is the input for the measurement, and TR is where the output of the saturation model or of the real actuator is connected. If the wind-up is not necessary, the terminal TR can be connected directly to the controller output OUT (see Fig. 6.13).

### 6.1.7 Saturation Model

The scheme of Fig. 6.11 can be implemented with op-amps. The only element that needs a comment is the saturation model. In fact, a simple op-amp with feedback is itself the saturation model because of the normal saturation level. However, if we want to simulate a well-defined saturation different from the ± power supply of the op-amp, the feedback shown in Fig. 6.14 can be used.

In the feedback we have a resistance and two Zener diodes connected with opposite polarization. The saturation level is the Zener diode voltage $Z$. If the absolute value of the model output is less than $Z$, the feedback is defined by the resistor. However, if the output reaches level $Z$, then one of the diodes conducts with voltage $Z$, and the other is open for the forward polarization. The effect is as if the feedback resistance be nearly zero, and the gain of the circuit drops. In other words, the output remains at level $Z$ (positive or negative). To get other saturation level with the same diodes, we can use the additional amplifier at the output, with variable gain.

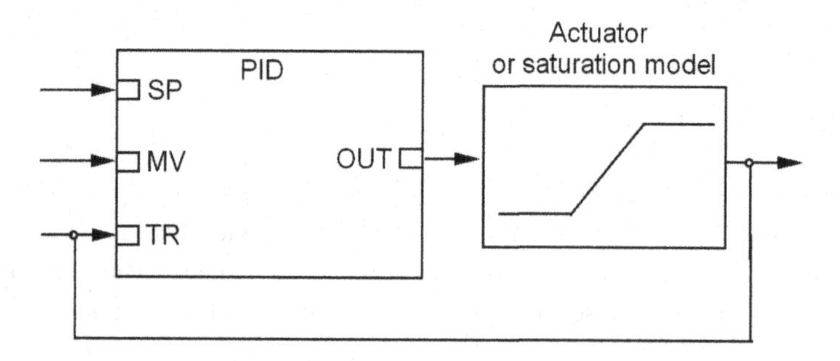


**Fig. 6.13** Anti-windup PID, terminals connections

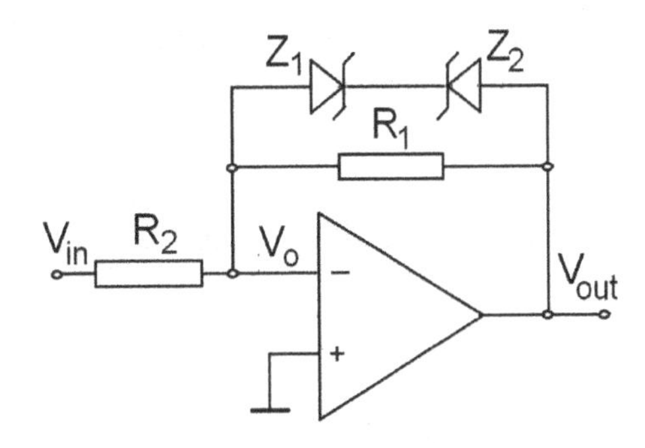


**Fig. 6.14** Saturation model

### 6.1.8 Direct Current Motor

In Fig. 6.15 we can see a scheme of a direct current motor with parallel configuration. The main parts of this motor are: The *armature (rotor), field windings brushes* and*commutator*. The field windings produce the magnetic flow $B$. This flow interacts with the magnetic field of the rotor that produces the torque $Q$ and makes the rotor move. The brushes and the commutator pass the rotor current to the rotor windings to keep the torque in the same direction, when the rotor moves. In the configuration in series, the rotor and the field windings are connected is series, so the same current $i$ passes through the windigs L and the rotor.

The rotor has a small internal resistance $r$. The torque depends on the field $B$ and the rotor current $i$, as follows.

$$Q = KiB \tag{6.13}$$

where $K$ is a constant. The angular velocity of the rotor abeys the following equation.

$$\frac{d\omega(t)}{dt} = \frac{Q(t)}{I}, \tag{6.14}$$

where $I$ is the moment of inertia of the rotor.

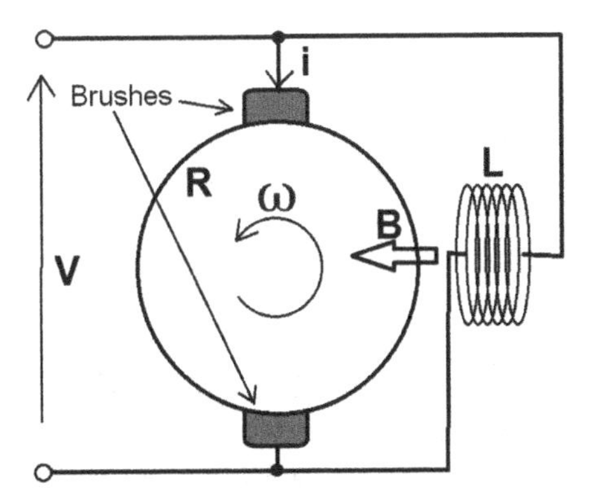


**Fig. 6.15** DC motor, parallel configuration

At the same time, the magnetic field of the rotor induces a *counter electromotive force* (back emf) $E$, opposite to the supply voltage $V$. This force depends on the rotor velocity $\omega$.

$$E = C\omega \tag{6.15}$$

The current $i$ is defined by the following equation.

$$i = \frac{V - E}{r} \tag{6.16}$$

From these equations we obtain:

$$\frac{d\omega(t)}{dt} = \frac{Ki(t)B}{I} - \frac{J(t)}{I} = \frac{K(V(t) - C\omega(t))B}{Ir} - \frac{J(t)}{I}, \tag{6.17}$$

where $J$ is the mechanic load. In terms of the Laplace transform we have

$$\omega(t)s = \frac{K(V(s) - C\omega(s))B}{Ir} - \frac{J(s)}{I}$$

ó

$$\omega\left(s + \frac{CB}{Ir}\right) = \frac{KV(s)}{Ir} - \frac{J(s)}{I}$$

Now, we can calculate the transfer function $V \to \omega$:

$$G_{V\to\omega} = \frac{H}{s + Z}, \tag{6.18}$$

where

$$H = \frac{K}{Ir}, \; Z = \frac{CB}{Ir} \tag{6.19}$$

The transfer function with respect to the load $J \to \omega$ is

$$G_{J\to\omega} = \frac{D}{s + Z}, \tag{6.20}$$

where

$$D = -\frac{1}{I}, \; Z = \frac{CB}{Ir}$$

The time constant of the motor response is equal to $\frac{Ir}{CB}$ for both transfer functions. Observe that

$$G_{J\to\omega} = \frac{D/Z}{s/Z + 1} = \frac{DIr/(CB)}{s/Z + 1}$$

This means that if $r \to 0$ then the velocity does not depend on the load. For small values of $r$, the influence of the load is small, and the velocity is approximately constant. This is the great advantage of the DC motors. Moreover, these motors can be easily controlled by the rotor voltage or by the field winding current.

## 6.2 Servomechanisms, Motor Control

The *servomechanism* is a device that control motion and position. This is a closed-loop control circuit with position-sensor feedback (Fig. 6.16).

### 6.2.1 Electrical Servomechanism, Rotative Movement

The transfer function of the DC motor has been discussed in previous Sect. 6.1.8, where we obtained the following.

$$G_{V\to\omega}(s) = \frac{H}{s+Z}, \tag{6.21}$$

where $H$ y $Z$ are constants that depend on the electro-mechanic parameters of the motor. Here, we are interested in the angle of movement $x$, rather than the velocity. For the angle, the transfer function is

$$G_{V\to x}(s) = \frac{H}{s(s+Z)} \tag{6.22}$$

We will start with a simple servomechanism, where the controller is of proportional action (P), and its gain, together with the power amplifier is equal to $K$. We also suppose that the angle sensor provides the ideal measurement, with gain equal to one.

The transfer function of our servomechanism will be

$$G_{a\to x}(s) = \frac{KH/(s(s+Z))}{1+KH/(s(s+Z))} = \frac{1}{s^2/(KH)+Z/(KH)s+1} \tag{6.23}$$

From (6.19) we have $Z = \frac{CB}{Ir}$, where $C$ y $B$ are constants. For big values of the product $Ir$ (moment of inertia and the rotor resistance), the term $Z/(KH)$ becomes small. This means that the movement of the servomechanism may have oscillatory character.

When the set-point (required angle) changes rapidly, or in the case of strong disturbances, the rotor current may increase too much in the transitory process. To avoid such problems, additional feedbacks are added to the control circuit, as shown in Fig. 6.17.

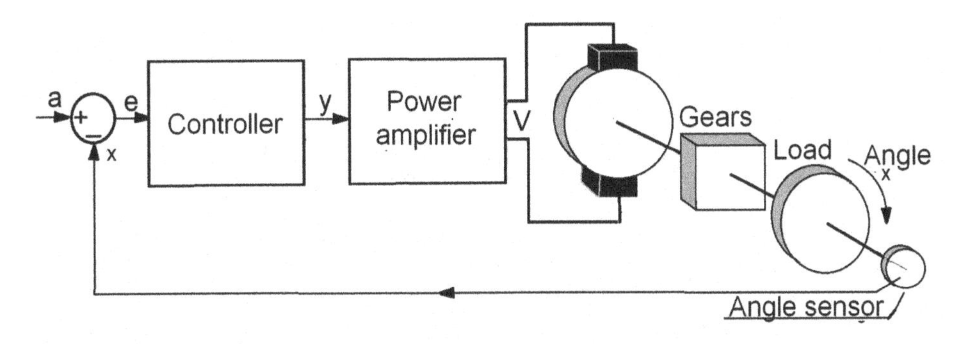


**Fig. 6.16** Servomecanismo con motor de corriente directa

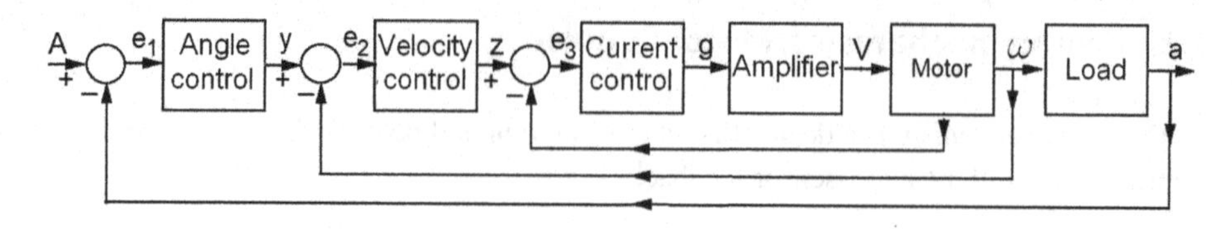


**Fig. 6.17** Feedback of angle, velocity and current. $a$ is the angle of movement, $\omega$ is the angular velocity

We will simulate the transient processes of a servomechanism with and without the internal current feedback. The controller of the main feedback is of type PI, and the current controller is of type P.

Model parameters are as follows.
Angle controller: Gain $K_r = 4$, duplication time $T_i = 1.5$
Current controller: Gain $K_c = 3$
Gear coefficient $K_{gear} = 0.5$
Parameter $C$ of the motor = 1
Rotor resistance $r = 4$[Ohm]
The corriente $i$ of the motor is given by the formulas (6.15), (6.16):

$$i = \frac{V - C\omega}{r} \tag{6.24}$$

This simulation uses a simple control circuit (Fig. 6.18), compared to the industrial solutions that may be more complicated, and also include the current limitation. In the model used here, the parameters are not optimal, to see the imperfections of the control and to make the response rather slow.

Figure 6.19 shows the comparison of the control with only one, angle feedback (plot 1), and the feedback of both the angle and the current. It can be seen that the current feedback improves the response of the mechanism. Figure 6.20 depicts the multiple response curves, where the gain of the current controller changes from 0.5 to 10. It can be seen that the big gain of current controller does not decreases the system stability.

The circuit of Fig. 6.18 can be constructed using the circuits discussed before. The PI controlled may be as in Fig. 6.4, and the circuit of Fig. 6.6 may be used as the power amplifier.

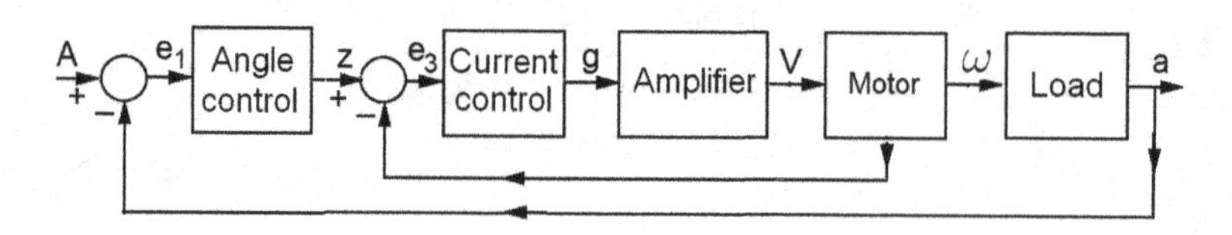


**Fig. 6.18** Angle and current feedback

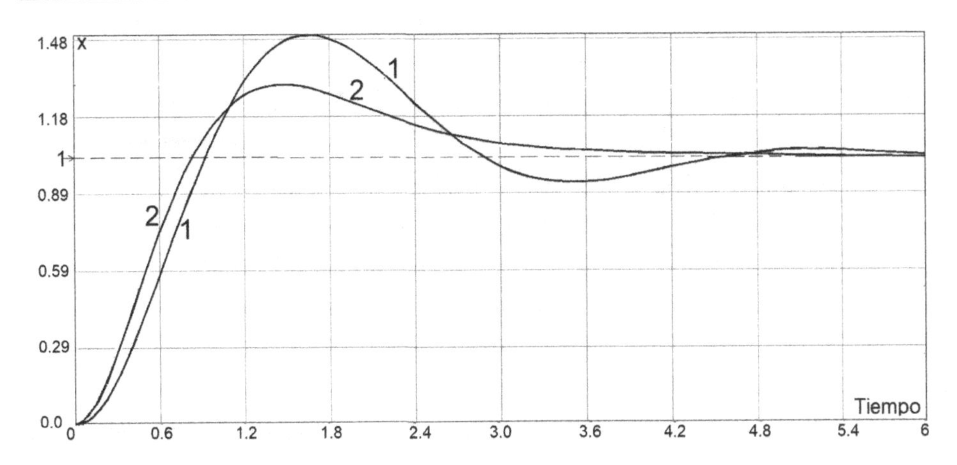


**Fig. 6.19** Angle feedback only (plot 1), and angle-current control (2)

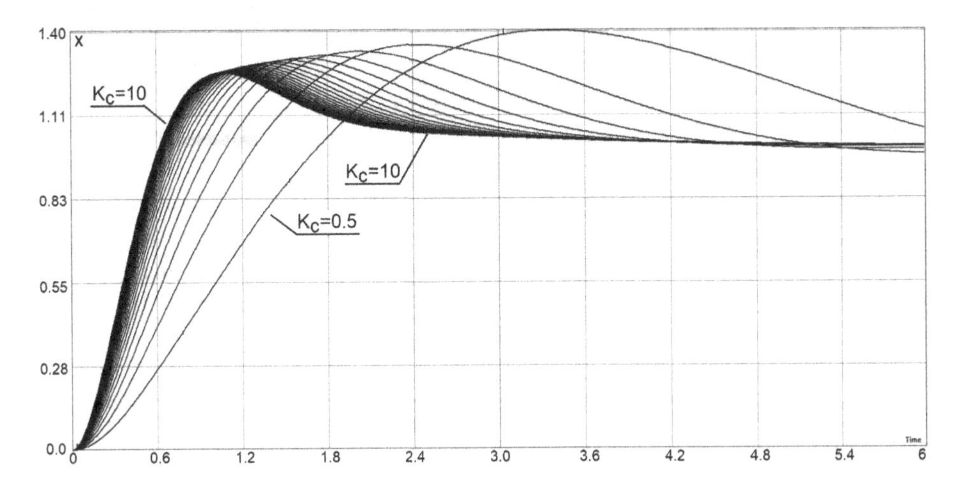


**Fig. 6.20** Angle and current control

## 6.3 Stepper Motor Control

The *stepper motor* is recently used in many electromechanical devices, instead of feedback-controlled servomechanisms (see Sect. 6.2). Stepper motor normally works in open-loop control circuits. This motor is controlled by electric pulses, each one of them making the motor move for one small step. This movement is done by relatively strong magnetic forces. So, it is supposed that, while receiving a pulse, the motor **must** move one step. With an appropriate pulse generator, we can move the motor by desired angle of rotation, without using a feedback mechanism.

The stepper motor cannot work without the corresponding electronic control circuit. It is an example of a mixed electronic-mechanical system, that belongs to the field recently referred to as *mechatronics*.

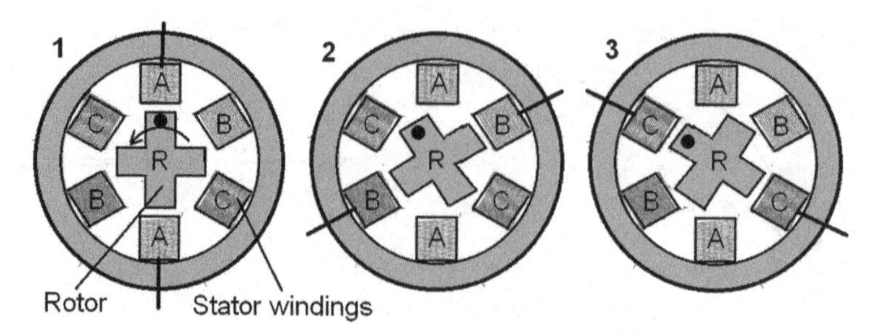


**Fig. 6.21** A stepper motor

Figure 6.21 shows the idea of the stepper motor. The two windings A-A are connected in series, and so are the windings B-B and C-C.

If the current flows through the windings A-A, the rotor is forced to take the position as shown in part 1 of the figure. If the current change to the windings B-B, the rotor moves by one step. Then, if the current is passed to windings C-C, we have the next step movement. This way, the motor change the rotating angle according to the number of received pulses.

This is a simple example of the motor with three pairs of poles. There are motors with multiple pole pairs that provide better precision of movement.

The stepper motor needs a driver that provides the power pulses to the windings, and the pre-driver, which is a pulse generator composed by a digital logic circuit are a microprocessor. In Fig. 6.22 there is an example of the driver that supplies the current to one motor winding. Note that the driver can provide the current in two opposite directions.

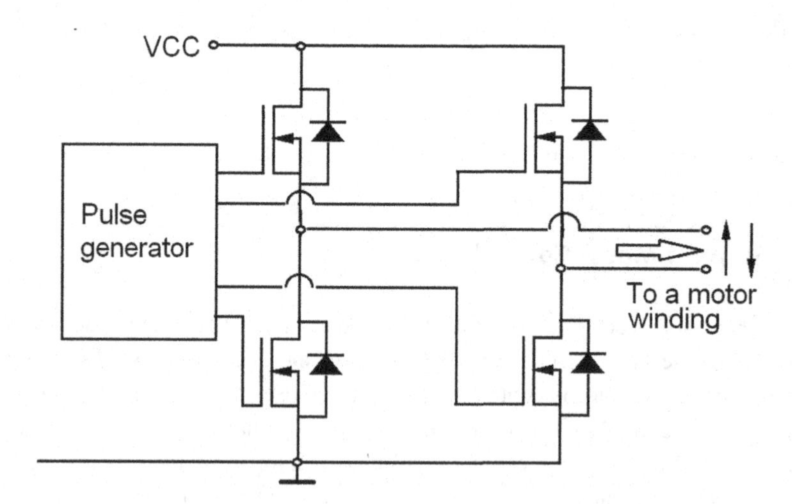


**Fig. 6.22** A stepper motor driver

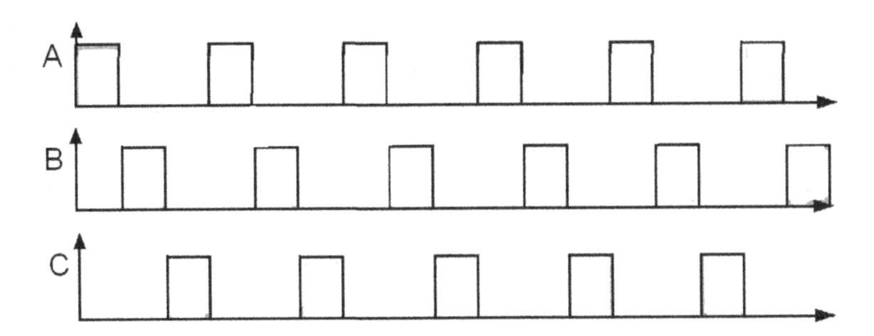


**Fig. 6.23** A stepper motor pulse train

The pulse generator should provide the pulse trains for the stator windings. Figure 6.23 shows an example the pulses needed for the windings A, B, and C, for a simple forward rotating. Depending on the motor construction, the pulses should have the appropriate polarization.

The pulse frequency defines the motor speed that can vary from zero to a high- speed rotation. This flexibility of the stepper motor makes it extremely useful in electro-mechanic applications.

# Digital Signal Processing, Control and Hybrid Circuits

7

## 7.1 Introduction

In automatic process control, the measurements of controlled variables do not have to be continuous. For example, in the slow processes like big tanks or heaters, the time-constants may reach hours or even weeks (as in blast furnaces). Such object may be observed in some discrete time instants with frequency of one per several seconds or one per hour, without significant loss of information. In automatic control, these systems are called *sampled data systems*. We will also refer to such object as *discrete-time systems*.

On the other hand, recently, the data can be collected with high velocities. For example, in audio signal processing, the sample frequency is greater than 40 kHz, with the sufficient velocity for sampled data processing. The data are received as sequences of pulses that come in discrete time instants. The sample frequency determines the possibility of retrieving the original, continuous signal from sampled data. There exists the minimal sample frequency that allows this, defined in the *theorem of Nyquist–Shannon*. The theorem states that, for the signal that contains the maximal frequency of $f$ Hz, the minimal necessary sample frequency is equal to $2f$ Hz.

## 7.2 Z-Transform

Z-transform is an important tool in sampled data analysis and signal processing. Here, we do not enter in all the theory of Z-transform. What the reader needs to know are basic and simple facts. There is a huge literature on the subject, see for example [4].

© The Author(s), under exclusive license to Springer Nature Switzerland AG 2023

S. Raczynski, *How Circuits Work*, Synthesis Lectures on Engineering, Science, and Technology, https://doi.org/10.1007/978-3-031-34934-8_7

Given a sequence of data $x_n$, $n = 1, 2, 3 \ldots$, the Z-transform is defined as below.

$$\mathcal{Z}\{(x_n\} = \sum_{n=-\infty}^{\infty} x_n z^{-n}, \tag{7.1}$$

where $z$ is a complex variable, and the data is taken in time instants $T, 2T, 3T, \ldots$. The transform (7.1) is called *bilateral*. If the sum is taken from 0 to $\infty$, then the transform is *unilateral*. Here, $z$ is a complex variable that can be represented as $z = Ae^{j\phi}$, where $A$ is a real number (amplitude), $j$ is the imaginary unit, and $\phi$ is the phase angle.

While compared with the Laplace transform (Sect. 3.1.2), it can be shown that there is a relation between the Laplace $s$ variable and $z$ variable, as follows :

$$z = e^{sT}, \tag{7.2}$$

where $T$ is the sampling period.

Consider a signal $y_n$, delayed by $k$ periods $T$ with respect to signal $x_n$: $y_n = x_{n-k}$. Suppose that the whole process starts with $n = 0$, $x_n = 0$ for all $n < 0$, and $m = n - k$. Thus, using (7.2), and the unilateral transform, we have

$$\begin{cases} \mathcal{Z}\{(y_n)\} = \sum_{n=0}^{\infty} y_n z^{-n} = \\ \sum_{n=0}^{\infty} x_{n-k} z^{-n} = \sum_{m=0}^{\infty} x_m z^{-m-k} = \\ z^{-k} \sum_{0}^{\infty} x_m \end{cases} \tag{7.3}$$

This means that the operation of delaying a signal by k periods T, it corresponds, in the domain of z, multiplying the corresponding transform by $z^{-k}$. In particular, one period delay means multiplying the transform by $z^{-1}$.

It should be noted that a sample data system is something essentially different from the continuous system. Using sampled-data we almost never suppose the period T approaching zero. It is always a finite system parameter. There are methods for converting sample data to continuous and vice versa like the zero-pole mapping, but these are always approximations, and the correspondence is is not unique.

One of the methods of finding a sampled data model that approximate a continuous one is to convert the differential equation model to difference equation and use the fact that one period delay is represented by $z^{-1}$.

Consider the following transfer function model.

$$G(s) = \frac{1}{a_n s^n + a_{n-1} s^{n-1} + \cdots + a_0} \tag{7.4}$$

We have $x(s) = G(s)u(s)$ where $u$ is the system input, and $x$ is the output. This imply the following equation.

$$a_n \frac{d^n x}{dt^n} + a_{n-1} \frac{d^{(n-1)} x}{dt^{(n-1)}} + a_{(n-2)} \frac{d^{(n-2)} x}{dt^{n-2}} + \cdots + a_0 x(t) = u(t) \tag{7.5}$$

Now, replace this equation with a difference equation. The derivative can be approximated by the expression $\frac{x_k - x_{k-1}}{T}$. Derivatives of higher order are approximated as follows.

$$\frac{\nabla_h^n x(t)}{T^n} = \frac{\sum_{i=0}^n (-1)^i \binom{n}{i} x(t - iT)}{T^n}$$

For example, an object of the first order with transfer function $G(s) = 1/(1 + \theta s)$ converts in the difference equation

$$x_k + \theta \frac{x_k - x_{k-1}}{T} = u_k,$$

($\theta$ is the time-constant of the object).

Now, recording that $z^{-1}$ means one period delay, we obtain

$$x(z) + \theta x(z) \frac{(1 - z^{-1})}{T} = u(z), \tag{7.6}$$

where $x(z)$ y $u(z)$ are z-transforms of $x$ and $u$, respectively. Thus, we have

$$G(z) = \frac{x(z)}{u(z)} = \frac{1}{1 + \frac{\theta}{T}(1 - z^{-1})} \tag{7.7}$$

The function $G(z)$ is the system transfer function in terms of z-transform.

In a similar way, we obtain the transfer function of an integrator $1/s$, as follows.

$$G(z) = \frac{1}{1 - z^{-1}}.$$

Consequently, the transfer function of a discrete-time PI controller is

$$G(z) = K \left( 1 + \frac{1}{T_i (1 - z^{-1})} \right) \tag{7.8}$$

As mentioned before, this correspondence between continuous and discrete-time systems is not unique. For example, if in the approximation of derivative we use the forward difference, obtenemos a different expression for the transfer function in $z$.

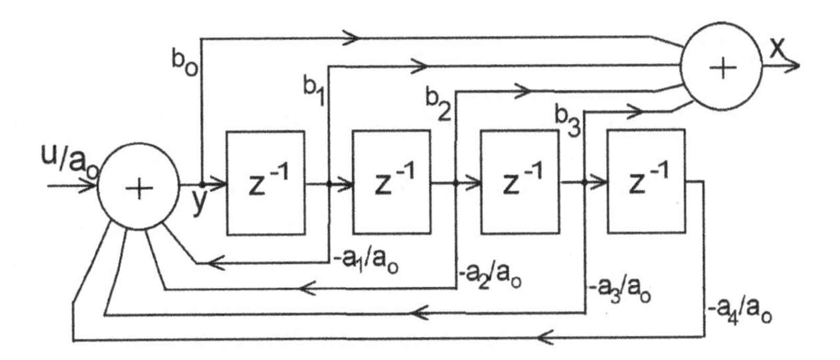


**Fig. 7.1** Circuit with transfer function (7.10)

For objects of higher order, the general expression for $G(z)$ has the form

$$G(z) = \frac{b_m z^{-m} + b_{m-1} z^{-m+1} + \cdots + b_0}{a_n z^{-n} + a_{n-1} z^{-n+1} + \cdots + a_0} \tag{7.9}$$

From this short introduction, we can conclude that the sampled data control systems may be implemented in hardware that uses elements of one-period delay. Such elements are easy to construct as hybrid devices, or to code in the corresponding software.

Consider, for example,

$$G(z) = \frac{x(z)}{u(z)} = \frac{b_3 z^{-3} + b_2 z^{-2} + b_1 z^{-1} + b_0}{a_4 z^{-4} + a_3 z^{-3} + a_2 z^{-2} + a_1 z^{-1} + a_0} \tag{7.10}$$

The links in this scheme (Fig. 7.1) have gains indicated over the arrows. The signal $y$, being the output of the sumator, is defined below.

$$y = \frac{1}{a_0} u - y \frac{a_4}{a_0} z^{-4} - y \frac{a_3}{a_0} z^{-3} - y \frac{a_1}{a_0} z^{-1} - y \frac{a_1}{a_0} z^{-1}$$

So, we have

$$y = \frac{u}{a_4 z^{-4} + a_3 z^{-3} + a_2 z^{-2} + a_1 z^{-1} + a_0}$$

From the scheme Fig. 7.1 we can see that

$$x = b_0 y + b_1 y z^{-1} + b_2 y z^{-2} + b_3 y z^{-3}$$

From these equations we can verify that the transfer function of the circuit is as in (7.10).

## 7.3 Aliasing

In Fig. 7.2 we can see the sampling data obtained from signals with different frequencies, f = 1 and f = 10. T is the sampling period. The sampling time instants are marked with "s". The signals are quite different from each other, but the obtained sampled data are the same. This ambiguity is called *aliasing*, and it is always present in sampled data systems. Moreover, there exist signals with other frequencies that may produce the same sequence of data.

One of the results of aliasing is that the frequency spectrum of a signal processing device repeats for frequencies that are multiple of the useful signal frequency. Sampling may also produce signals with frequencies that do not exist in the processed input signal (Fig. 7.3). This may produce an unnecessary noise and confusion in data processing.

While creating digital filters, these higher frequencies can be eliminated, using the first or second order analog low-pass filter with cut-off frequency slightly greater than the maximal frequency of the useful signal.

## 7.4 The Sample-and-Hold Element

The basic sampled data is a series of pulses with value proportional to the sampled continuous data. These pulses, in the ideal, theoretical approach are represented as Dirac's $\delta$ function. However, in real system, we rather need a finite, rectangular pulses that represent the data. The *sample-and-hold* (SAH) element provides such pulses, each of them with duration of one sampling period. The retained value is equal to the value of the continuous signal, taken at the sampling time instant, see Fig. 7.4. In the figure, the continuous signal is $x(t)$ and the sampled (measured) data are marked as $x_m(k)$.

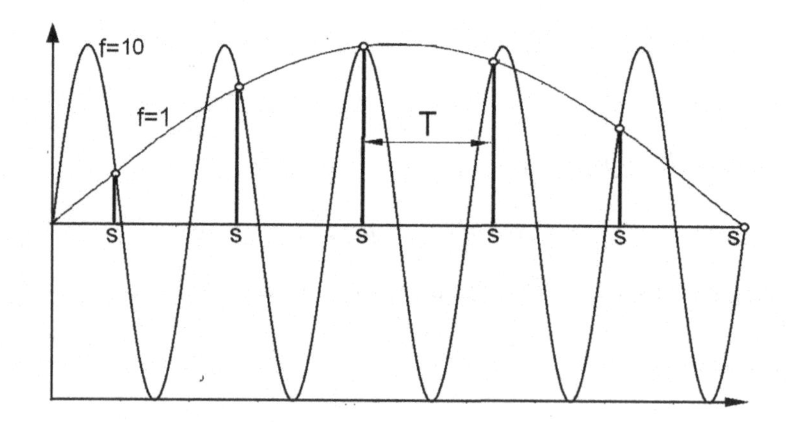


**Fig. 7.2** Sampling signals with different frequencies

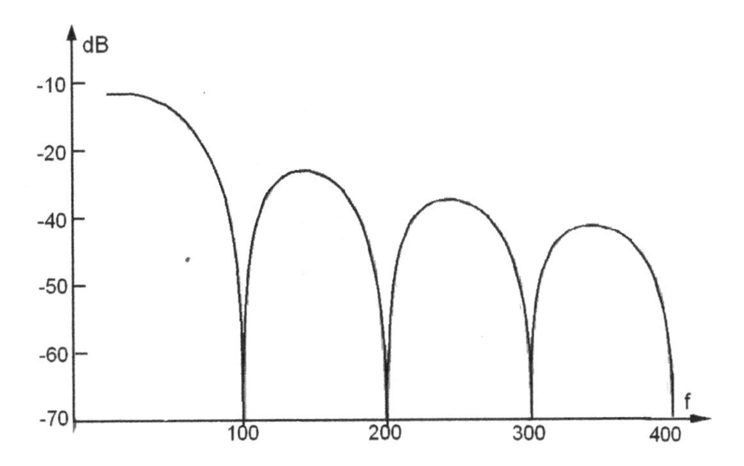


**Fig. 7.3** Repeating spectrum of a digital filter

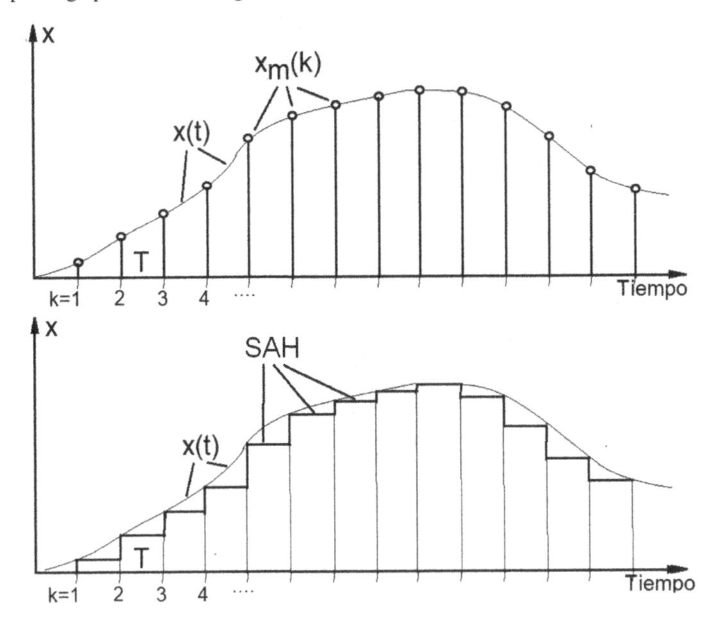


**Fig. 7.4** Sampled data and sample-and-hold

Figure 7.5 shows a possible electronic implementation of SAH. Both op-amps work as followers. The JFET transistor is a switch. It receives pulses in sampling moments that open JFET for a short time. The continuous signal is taken from the output of the first op-amp. This value, in the sampling time instant, charges the capacitor C. Then, the capacitor voltage is passed to the output through the second follower. Note that the followers are necessary. The first one is used to avoid the big charging current of the capacitor to be taken from the signal source. As for the second follower, observe that we can hardly take directly the

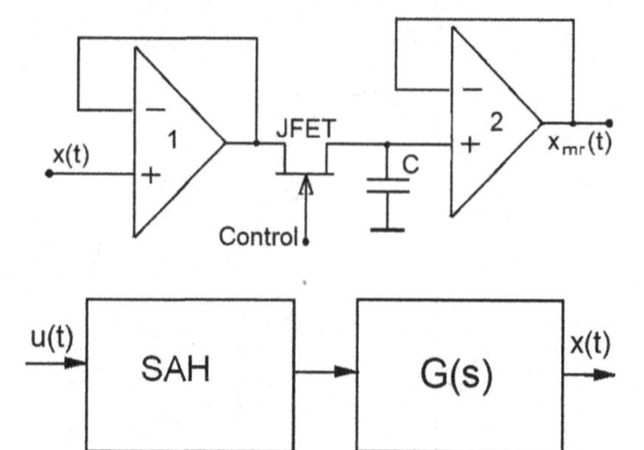


**Fig. 7.5** Sample-and-hold circuit

**Fig. 7.6** Sample-and-hold and a continuous objects

voltage from the capacitor because the load resistance is unknown and the load can discharge the capacitor. If the second op-amp has the FET input stage, practically no current is taken from the capacitor, but the final load may receive a considerable current.

The transfer function of the sample-and-hold element is as follows.

$$V(s) = \frac{1 - e^{-sT}}{sT} \tag{7.11}$$

Observe that if the object of Eq. (7.11) receives a Dirac pulse, then the part $1/(TS)$ integrates it and provides output increment proportional to the pulse. The part $-e^{-sT}/(sT)$ integrates the previous pulse and subtracts it from the output. So, the output increment is proportional to the difference between the two consecutive pulses that give us the required SAH response.

Now, consider a connection of a SAH and a continuous object (Fig. 7.6).

The transfer function of this connection is

$$F(z) = (1 - z^{-}1)\mathcal{Z}\left[\mathcal{L}^{-}1\left\{\frac{G(s)}{s}\right\}\right]$$

where $\mathcal{Z}$ is Z-transform, and $\mathcal{L}^{-1}$ is the inverse Laplace transform.

Now, let us see the influence of the sampling time on system stability. We consider a very simple example, where the only sampled data element is the measurement instrument that takes a measure and retains it during the sampling period T (Fig. 7.7). In the real industrial applications there are many possible configurations of continuous and discrete-time objects, and the SAH can appear in various places in the block diagram. The example below has only one SAH, to illustrate the influence of the sampling period.

As can be expected, for small T, the system behavior is similar to the continuous version.

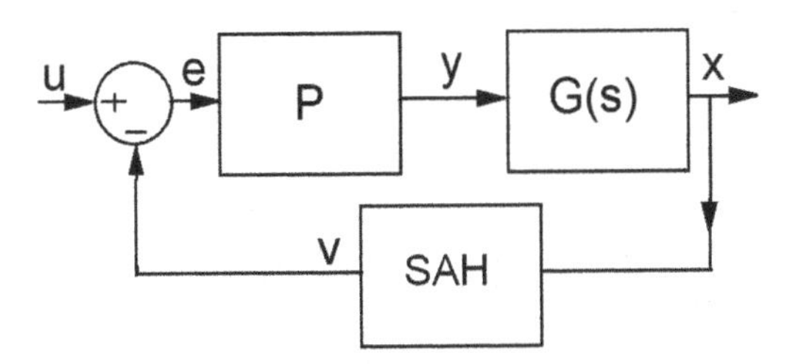


**Fig. 7.7** Control circuit with sample-and-hold

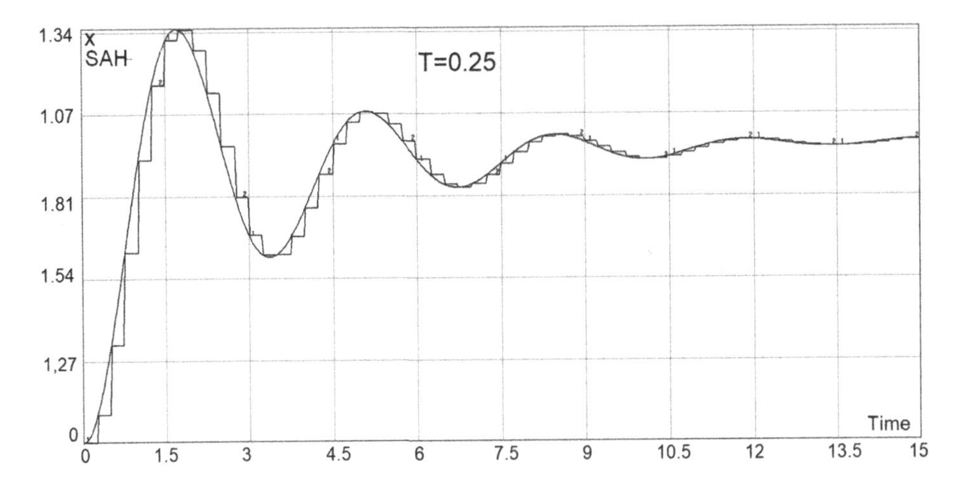


**Fig. 7.8** System response, T = 0.25

System parameters are as follows

Controller gain $K_r = 3$

Integration time $T_i = 6$

The process: $G(s) = \dfrac{1}{s^2 + 1.5s + 1}$

In Figs. 7.8, 7.9 and 7.10 there are system responses to the unit step input, with T = 0.25, 0.5 and 0.9, respectively. It can be seen that for the small T the response is oscillatory, but stable. For T = 0.5 the oscillations are greater, but the system is still stable. However, if T = 0.9, the system becomes unstable and the oscillations grow.

## 7.5 Some Hybrid and Time-Delay Circuits

By a *hybrid* circuit, we mean an electronic device that includes the time-discrete and logic elements like AND, OR, NOR gates and switches, as well as analog, continuous devices like integrators, inertial objects and other continuous elements.

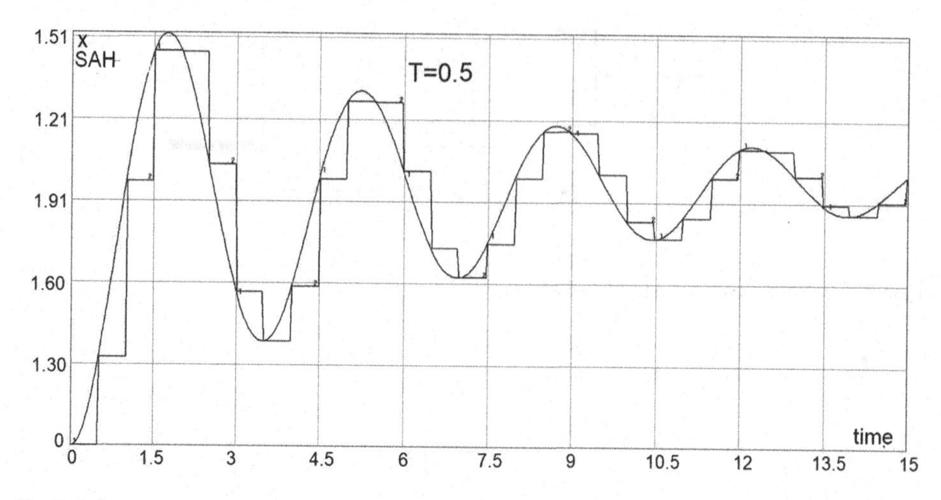


**Fig. 7.9** System response, T = 0.5

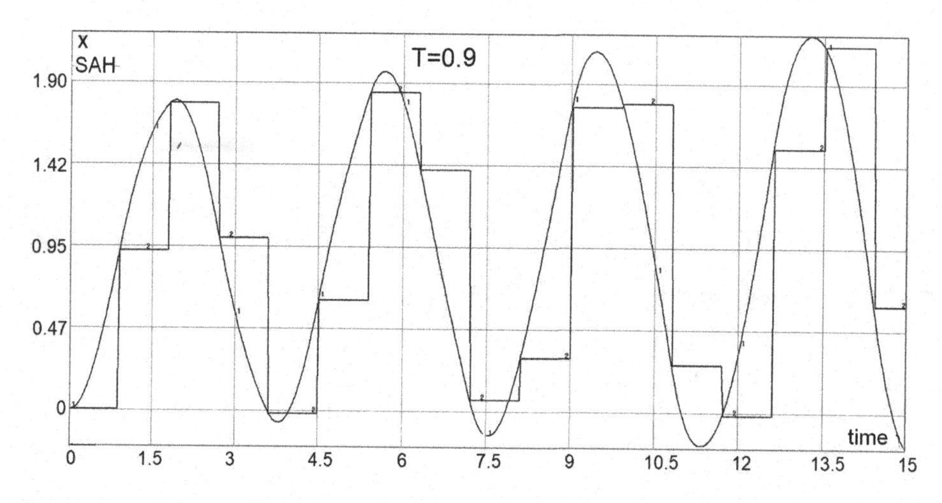


**Fig. 7.10** System response, T = 0.9

The sample-and-hold circuit discussed in Sect. 7.4, Fig. 7.5 is a hybrid device, where the JFET works as logical element (a switch).

### 7.5.1 Analog-to-Digital Convertor

The *analog-to-digital* (AD) converter receives a continuous analog signal and converts it in the digital, mostly binary, representation. The simplest AD converter is just a comparator. Depending on the difference between input voltages, it provides logical "0" or "1" out-

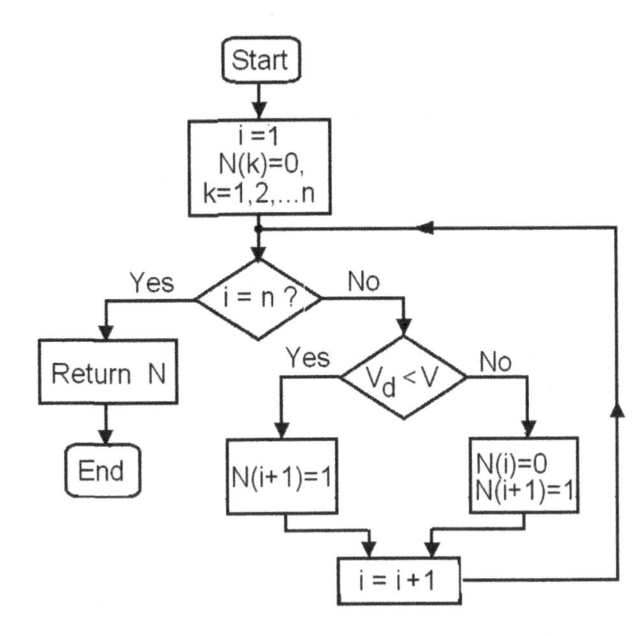


**Fig. 7.11** Analog-to-digital algorithm

put. The conversion to more precise digital form is somewhat more complicated. The AD algorithm is shown in Fig. 7.11.

In this algorithm, n is the required number of binary positions of the result, i is the iteration number. The input analog voltage is $0 < V < 1$. The purpose is to get a binary representation N(1), N(2), …, N(n) of $V$, N(i) = 0 or 1. The voltage $V_d$ is actual value that corresponds to N,

$$V_d = N(1)/2 + N(2)/4 + N(3)/8 \ldots$$

In the diagram Fig. 7.11, i is the iteration number.

Example: Let us convert voltage $V = 0.7$ in the binary form, in 4 iterations. The consecutive iterations are as follows.

$\mathrm{i} = 1, \mathrm{N} = 0.0000, V_d = 0,\ V_d < V, \rightarrow \mathrm{N}(1) = 1, \mathrm{i} = 2$
$\mathrm{i} = 2.\ \mathrm{N} = 0.1000, V_d = 0.5,\ V_d < V, \rightarrow \mathrm{N}(2) = 1, \mathrm{i} = 3$
$\mathrm{i} = 3.\ \mathrm{N} = 0.1100, V_d = 0.75,\ V_d > V, \rightarrow \mathrm{N}(2) = 0, \mathrm{N}(3) = 1, \mathrm{i} = 4$
$\mathrm{i} = 4.\ \mathrm{N} = 0.1010, V_d = 0.625,\ V_d < V, \rightarrow \mathrm{N}(4) = 1, \mathrm{i} = 5, \text{Stop}, \mathrm{N} = 0.1011$

The final value of $V_d$ is 0.6875. It can be seen that increasing number of iterations, the register N tends to the binary representation of 0.7.

A standard resolution of the AD integrated circuits is 12 bits. However, for more quality devices, the resolution can be of 16 or more bits.

## 7.6 Delay Lines

### 7.6.1 Long Line Delay

While sending data like voice signals, through long transmission lines, it can be observed the delay in the received signal. This is the phenomenon that is always present in transmission cables. Recall two facts. First, any wire we use has not only resistance, but also inductance because it generates a magnetic field around. Second, there is always certain capacitance between two conductors. Thus, in the long transmission line (Fig. 7.12), these two parameters are distributed along the line. The model of such long line is not very simple. The propagation of the signal is described by a partial differential equation that will not be discussed here.

### 7.6.2 LC Delay Circuit

The model of a long delay line can be simplified, dividing the line into a finite number of inductances and capacitances. Such lines can be realized with discrete elements, when the exactness of approximation depends on the number of used elements. In practical applications, such lines can be used to obtain delays, without using long cables.

Figure 7.13 shows an example of such model, with only five LC elements. This is a rough approximation of the exact signal delay device. However, even is such model we can observe the delay effect. Figure 7.14 depicts the results of a simulation of the line of Fig. 7.13. In the figure, we can see the plots of the voltages $V_1, V_2, \ldots, V_5$, while the input voltage was a unit step function, $L = 10\,\mu H$, $C = 0.5\,\mu F$. The boldface line $V_5$ is the output of the model. It can be seen that the signal has a nearly zero response for the delay time, and then grows rapidly.

Such delay devices, with 10 or more elements has been used to achieve various audio effects. Already in 1939, this circuit has been implemented by Laurens Hammond in his wonderful invention, known as the Hammond organ. In the organ that, by the way, contained few electronic tubes, the generated signal is passed to the 18-stages LC delay line. There is a "Hammond scanner" that consists in a capacitor, with one rotating part, and the other composed by several separated elements. These elements are connected to the consecutive

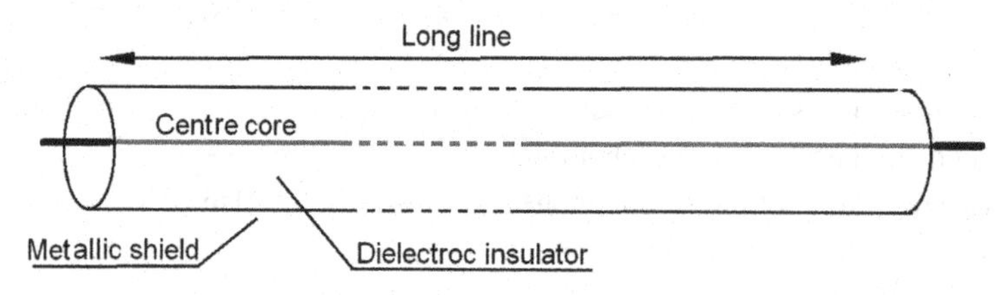


**Fig. 7.12** Long transmission line

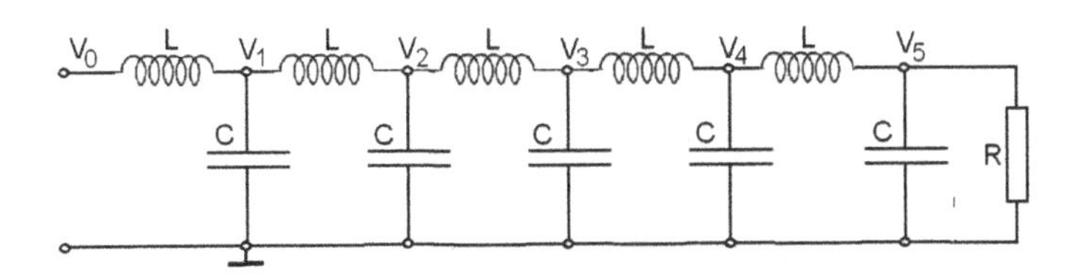


**Fig. 7.13** LC delay line

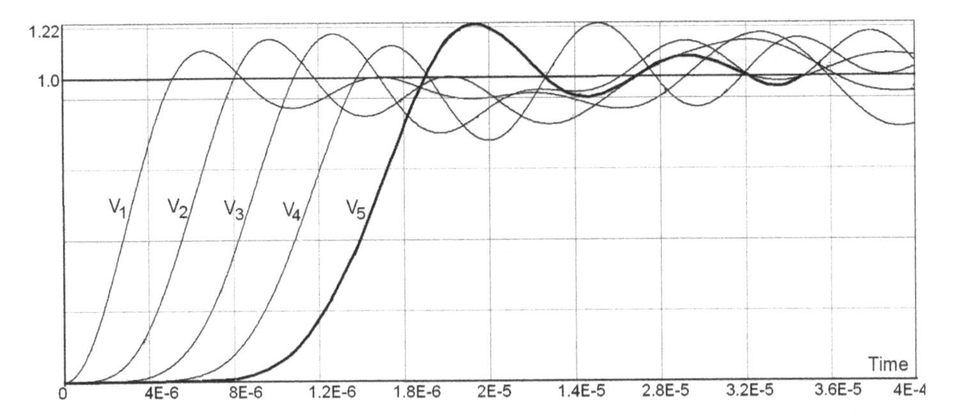


**Fig. 7.14** Simulation of LC delay line

points of the delay line, so, while rotating, the scanner provides signal collected from the line, going forwards and backwards. This results in a "Doppler effect" that produces a famous Hammond vibrato, never surpassed by other digital or hybrid systems of recent years.

In Fig. 7.15 there is a scheme of the Hammonf organ delay line with the scanner (S). It is a capacitor with one plate (P) rotating and the other plate, divided in several parts that are connected to various points of the delay line. The number of LC stages of the line may vary and some other modifications may be added, but this is the main idea of the Hammond vibrato.

### 7.6.3 Bucket Brigade Delay (BBD) Line

The circuit name was chosen because the BBD works similar to a line of people passing buckets of water or other things. It was developed by the Matsushita Company in mid-70s

The idea of the BBD circuit is to pass the electric charge from one capacitor to another, as shown in Fig. 7.16.

The BBD delay consists of an array of capacitors that may receive the charge from the followers marked with $f$. There are two sets of switches> $a$ and $b$. When switches $a$ close, then the capacitor $C_1$ receives the the charge from the input V. Capacitors $C_3.C_5, \ldots$ receive

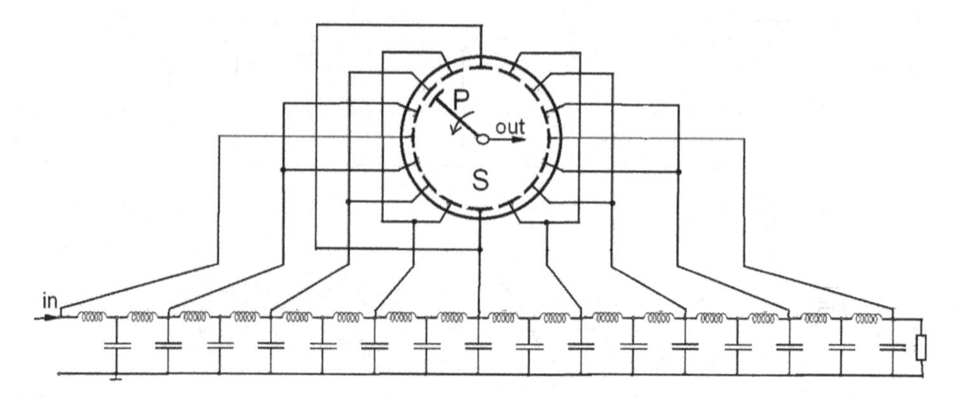


**Fig. 7.15** Hammond delay line and scanner

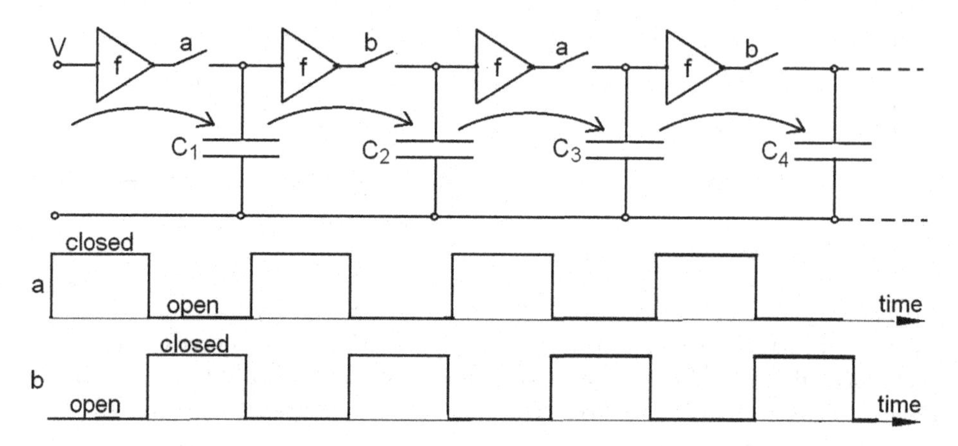


**Fig. 7.16** BBD circuit

the charge from $C_2$, $C_4.C_6, \ldots$, respectively, so the voltage of $C_1$ becomes equal to V. When switches $a$ open, and $b$ close, then $C_2$, $C_4.C_6, \ldots$ receive the charge from $C_1$, $C_3.C_5, \ldots$, respectively. The capacitor $C_2$ is charged to the voltage V. If the cycle repeats, the voltage V is passed to $C_3$, etc. This way, the input voltage travels along the line, and is received at the final stage with certain delay.

Note that the followers are necessary. Without the separating follower, the passed voltage would decrease at each switching cycle. The state of the switches is controlled by a rectangular sequence of clock pulses. The clocks are phase shifted (inverted), so, the switches $b$ be open when the switches $a$ are closed and vice versa.

Figure 7.17 shows a practical implementation of BBD delay, using MOSFET transistors.

In the audio applications, BBD delay with feedback produces the *reverb* or *echo* effect. First BBD integrated circuits had 512 or 1024 stages. Recently, the most used BBD use 4096 MOSFET-capacitance stages, that may produce a good quality effects.

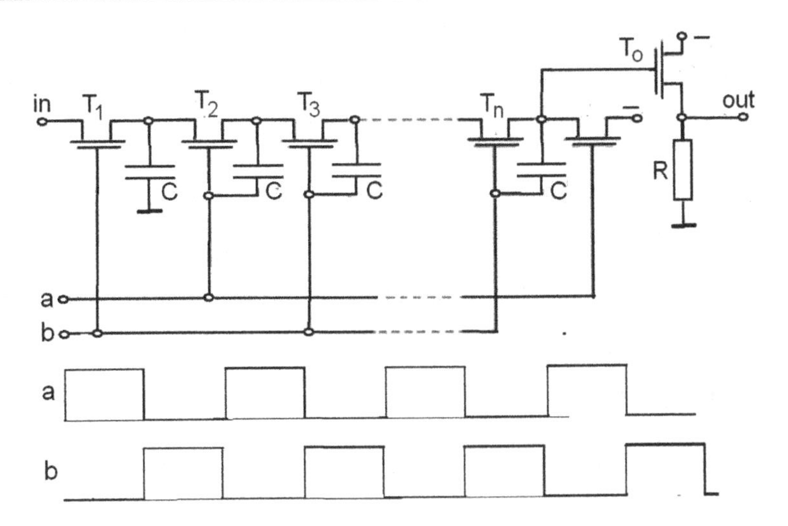


**Fig. 7.17** BBD circuit with MOSFET

Total delay of the BBD with clock frequency $F$ is equal to $D = N/(2F)$, where N is the number of steps. For example, with $F = 13000\,\text{Hz}$ and $N = 4096$, we get the delay of 0.157 s that provides a good echo effect. Lower clock frequencies can give us greater delays, but then, the aliasing between the clock and the voice signal may occur, producing noise or other, unwanted frequencies on the output.

The clock that generates controlling pulses can have constant or variable frequency. If we modulate this frequency, for example, between 2 and 10 Hz, then the delay changes, producing the Doppler effect similar to that of the Hammond LC line with scanner. This permits to design circuits that imitate the original Hammond vibrato.

### 7.6.4 CMOS Digital Delay

Other way to implement the delay in audio applications is the *digital delay*. These devices are digital microprocessors that are pre-programmed to provide delayed signal. The are based on the CMOS technology, and equipped with AD and DA converters and anti-aliasing filters. This kind of circuit offers a high quality sound processing, with delay of 300 msec and more.

The digital microprocessors are not the topic of this book, so we will not discuss these circuits. For more detail, consul the datasheets of the available ICs, for example the PT2399 IC of Princeton Technology Corporation.

## 7.7 Sound Effects

The delay line circuit is the basic element of several sound effect devices. Adding a feedback to the delay line circuit we can obtain a reverb or echo effect. In general, the audio signal with no effect is called *dry* signal, and adding an effect we get a *wet sound*. The share of dry/wet can be controlled by the user.

As mentioned before, if we use a controlling clock that can be frequency-modulated, the delay may oscillate, adding the phase modulation to the signal. This may provide a tremolo effect. The delay circuits are used to achieve other effects, like the following.

**Comb filtering**. In this effect, a copy of the original signal is created, and slightly delayed, up to 10 ms. The interference of this signal with the original one results in amplifying or eliminating some frequencies ("comb filtering" or destructive interference).

**Chorus effect** consists in generating, out of the original input, signals that have slightly different pitch and timing. This is like multiple musicians or vocalists play the same melody, with small differences. The superposition of the multiple sounds give us the chorus effect. Unless the vibrato effect, the chorus uses a subtle changes in pitch and phase, following a low frequency oscillator that controls the delay line.

The chorus *rate* parameter controls the speed of modulation, normally between 0.1 and 0.5 Hz. The *depth* defines the amount of the pitch modulation. In some implementation, the user can also add the feedback to the signal. The stereo spread of the device is controlled by the *width* parameter. With zero width, the sound is centered, with no stereo effect. Greater width give us stronger stereo sound.

**Flanger** effect works in high acoustic frequency range. The flanger action is similar to the chorus, but is uses shorter delays, no more than 30 ms. Flanger creates a copy of the original signal and makes it interfere with the original. This results in the comb filtering with multiple notches. Now, the amount of delay is modulated by a low frequency oscillator, and notch frequencies oscillate. This produces the "metallic" flanger effect.

**Phaser**. This circuit produces the phase-shift of certain frequencies. This signal is combined with the original. The effect is similar to the flanger. However, the resulting sound is not so sharp and strong as provided by the flanger.

**Overdrive**. Not all sound effects need delay lines. In early 1950s, the vacuum-tube amplifiers were not very powerful, and the musicians, mainly guitar players, have been frequently making de the amplifiers saturate, producing strong signal distortion. They observed that this kind of distortion adds a useful effect to the sound. The simplest overdrive consists in passing the strongly pre-amplified signal through a limiter. This converts the original vibration into the square wave signal, that may contain even more harmonics than the signal without distortion. The effect is commonly used in rock, blues and popular music.

**Wah-wah** effect consists in sweeping the peak response of a frequency filter up and down. The resulting sound has no non-linear distortion or delay, but has the frequency response changing in time. This provides a somewhat more natural and vibrating sound effect.

# Miscellaneous: Linear IC Applications, Logical Gates

8

## 8.1 Linear ICs, Op-amps

The op-amps has been discussed in several chapters, mainly Chaps. 2, 3, 6. Recall that the op-amp is a DC and AC component amplifier with very big gain and differential input (Fig. 8.1). The input impedance is more than 300 kΩ, and may be considered infinite for op-amps with input stage based on FET transistors. The following circuits may be constructed using the standard LM741 op-amp, or a JFET TL081.

Adding a negative feedback to an op-amp, we can get a great variety of circuits like filters, followers, inverters, sumators, integrators or differentiating circuits. Figure 8.2 shows the basic feedback op-amp circuit. The total transfer function of the circuit is

$$G(s) = \frac{x(s)}{u(s)} = -\frac{Z_1(s)}{Z_2(s)}. \tag{8.1}$$

Next, we will mention more practical applications of op-amps.

### 8.1.1 Stereo Mixer

Figure 8.3 presents the diagram of a simple audio mixer. The purpose is to construct a mixer with three input mono channels $in_1, in_2, in_3$, mix them and distribute between the left and right stereo outputs.

At the input we have three potenciometers $P_1$, $P_2$ and $P_3$ that control the gain for in1, in2 and in3. The three signals are amplified by non-inverting, high input impedance amplifiers. The capacitors at input are used to reject a DC component of input signals, if any. These should be a good quality capacitors, preferably ceramic.

© The Author(s), under exclusive license to Springer Nature Switzerland AG 2023

S. Raczynski, *How Circuits Work*, Synthesis Lectures on Engineering, Science, and Technology, https://doi.org/10.1007/978-3-031-34934-8_8

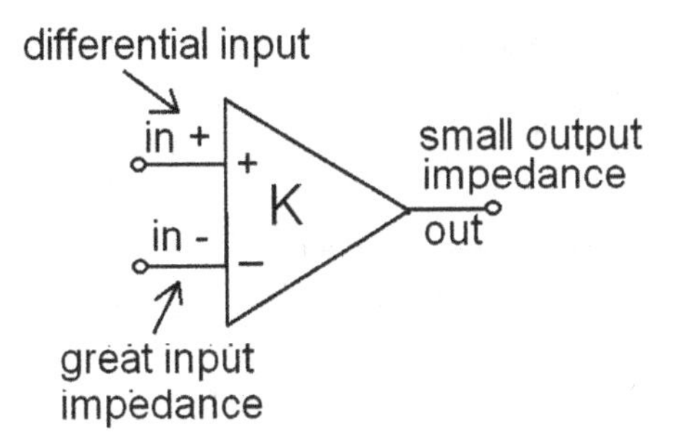


**Fig. 8.1** Operational amplifier

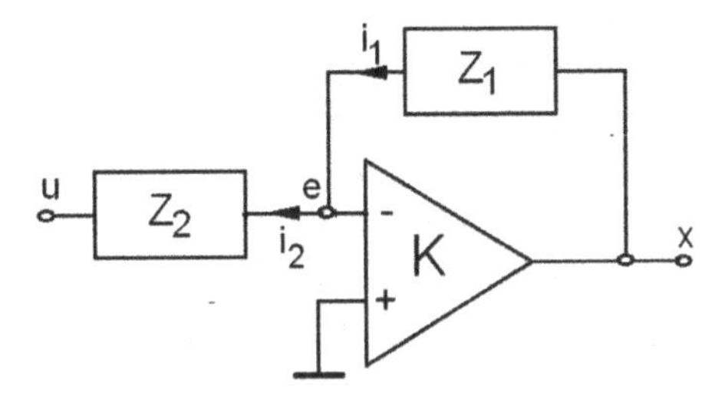


**Fig. 8.2** Operational amplifier with feedback

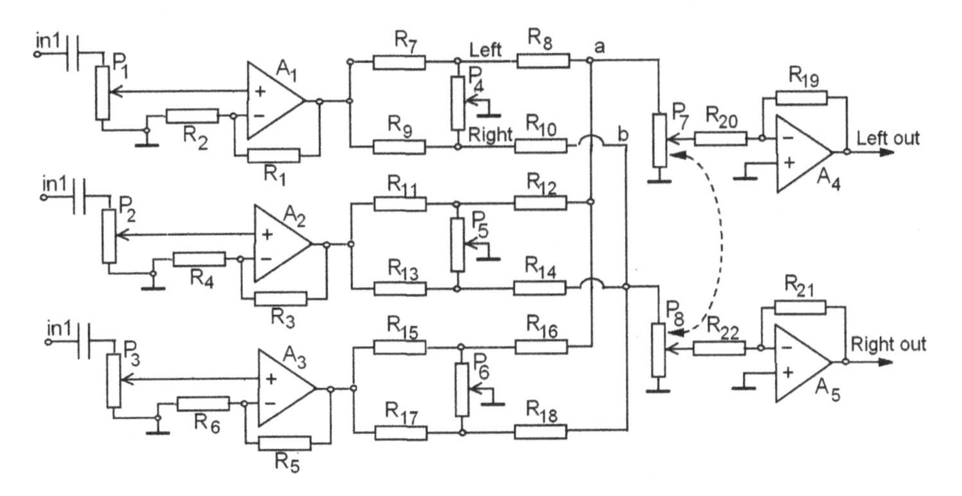


**Fig. 8.3** A stereo mixer

The amplified signal enters a passive red that distributes it between left and right stereo outputs (PAN feature). For example, in the channel $in_1$ we have resistors $R_7$, $R_8$, $R_9$ and $R_{10}$ and the potenciometer $P_4$. If the potenciometer is in high position, then the signal that goes to the left stereo output is blocked, and the signal is directed to the right channel. For the low position of $P_4$ the signal goes to the left channel. If $P_4$ is in middle position, both left and right channels receive the signal of in1. For the inputs in2 and in3, the PAN works in the same way.

The left channel is collected through the resistors $R_8$, $R_{12}$ and $R_{16}$ and goes to the output amplifier. In the similar way, $R_{10}$, $R_{14}$ and $R_{18}$ provide the right channel signal. The potenciometers $P_4$, $P_5$ and $P_6$ must have linear characteristics. $P_7$ and $P_8$ should be parts of a dual, stereo potenciometer. It will control the "master" volume.

This is the very simple mixer that can be improved and equipped with new features. For example, note that the position of any potenciometer $P_4$, $P_5$ and $P_6$, affects the signal in other channels. To avoid this, the resistors $R_8$, $R_{10}$, $R_{12}$, $R_{14}$, $R_{16}$ and $R_{18}$ may be replaced by the op-amp followers. If the mixer receives signals from balanced microphone cables, then the inputs should be modified to support the differential signals. The best way to do this, is to use an audio linear amplifier, like the LM386 of Texas Instruments. This IC needs only one, positive power supply and has the symmetric differential input. It needs no feedback, and offers the gain between 20 and 200 times, depending on configuration. If we use balanced inputs, it would be useful to, optionally, connect the "phantom power" DC supply of 9–50 V directly to the microphone inputs, through the resistances of 6.8kOhm.

As for the number of input channels, of course the mixer can have and number of inputs, repeating the input parts and the PAN network.

Other enhancement may include an equalizers (bass and treble), added at the output of $A_1$, $A_2$ and $A_3$ (see Sect. 3.2.6).

The mixer may also have the "send" and "receive" terminals that allow to use external effects like reverb or echo. The "send" may consist of new output terminals connected to points "a" and "b", and the additional stereo input to be mixed with the output stereo channels.

### 8.1.2 Doorbell Sound Generator

Figure 8.4 shows a simple doorbell circuit. There is an op-amp with negative feedback that is a twin T notch circuit, discussed before in Sect. 3.2.5. TT-circuit changes the phase of the signal by 180° near the notch frequency, making the feedback positive for that frequency. This makes the circuits oscillate. With appropriate values of the TT-circuit parameters, there are damped oscillations that produce a bell-like sound. PB is a momentary contact "spst" (normally open) push button. It provides a voltage pulse to the op-amp input that initiates the oscillation. The three potentionmeters define the frequency and the damping.

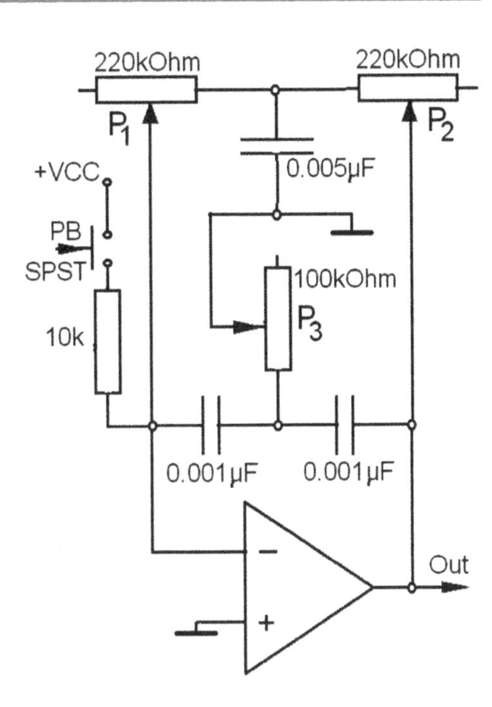


**Fig. 8.4** Bell sound generator

## 8.2 Logic Gates and Digital Circuits

Logic circuits manage inputs and outputs as logical values, treated as "0" (false) and "1" (true). In the physical implementation, these are "low voltage" (for "0") and "high voltage" (for "1"). In fact, a single transistor in common emitter configuration and collector resistor is a logical gate that provides the low response to high input and vice versa. Such circuit is a logical NOT gate that produces the negation of input logical value.

The main techniques used in the production of logic gates and corresponding integrated circuits are *Complimentary Metal-oxide Semiconductor* (CMOS) and *Transistor-Transistor Logic* (TTL) (see Sect. 1.10.5).

CMOS circuit uses a field-effect transistors of type NMOS or PMOS. These are more sensitive and more likely to get damaged easily as compared to the TTL The power consumption of CMOS chips is considerably lower than TTL. The supply voltage for TTL is approximately 5 V, while CMOS may need more than 10 V supply. CMOS gates may have multiple inputs, up to 12–14, while for the TTL number of inputs can hardly be greater that 10. The design of CMOS gates is simpler than for the TTL, which needs additional elements, like resistors. The signal propagation in CMOS gates is more than 20 ns, while in TTL it can be less than 10 ns.

In the following, we will concentrate rather on the logical functions than on a specific fabrication techniques.

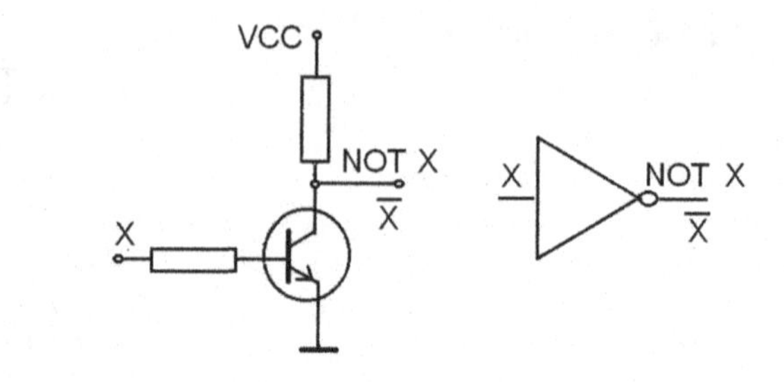


**Fig. 8.5** NOT gate

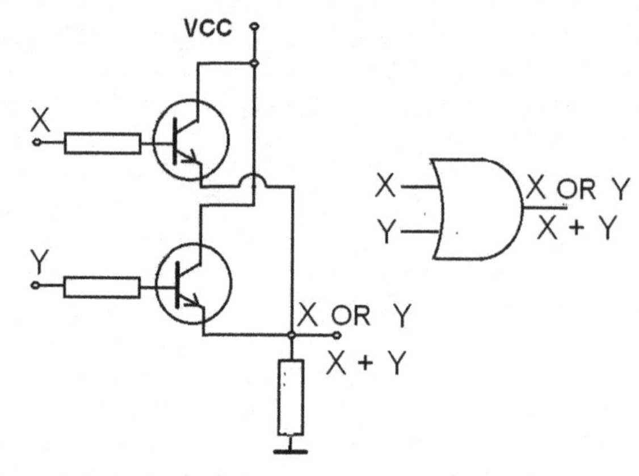


**Fig. 8.6** Logic OR gate

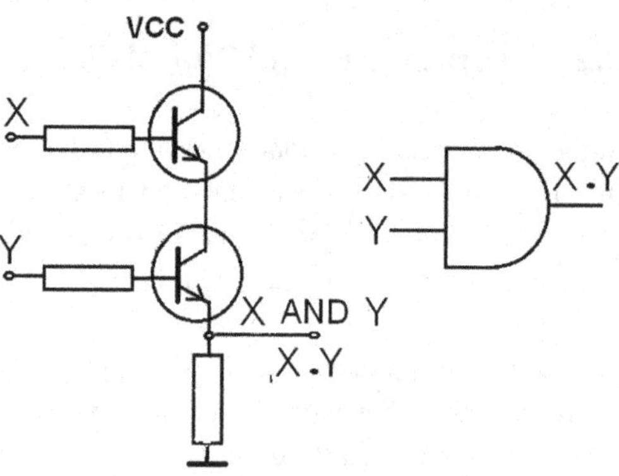


**Fig. 8.7** Logic AND gate

Figure 8.5 represents a simple implementation of the logic NOT operation. The circle on the symbol, called a bubble is used to indicate a logic negation. The figure shows an example of implementation with transistors, and the corresponding symbol of the NOT gate.

Figures 8.6, 8.7 ,8.8 and 8.9 show the implementations and symbols to the OR, AND, NOR (NOT OR) and NAND (NOT AND) gates, respectively.

A frequently used is an *exclusive OR gate* that realizes the operation

$$Z = X \cdot \bar{Y} + \bar{X} \cdot Y = (X + Y) \cdot (\bar{X} + \bar{Y})$$

Figure 8.10 depicts a possible construction of the XOR gate, using NAND gates.

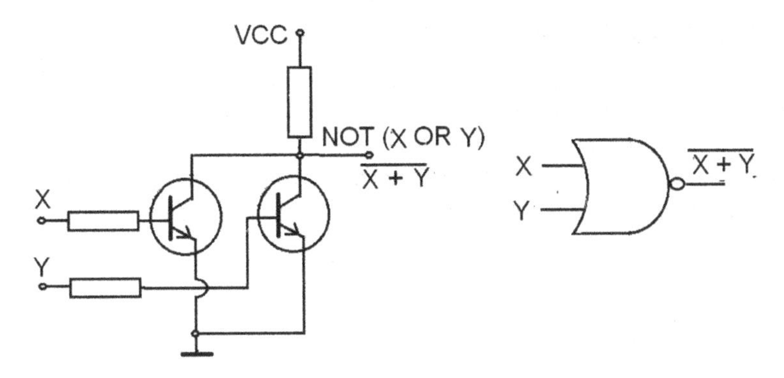


**Fig. 8.8** Logic NOR gate

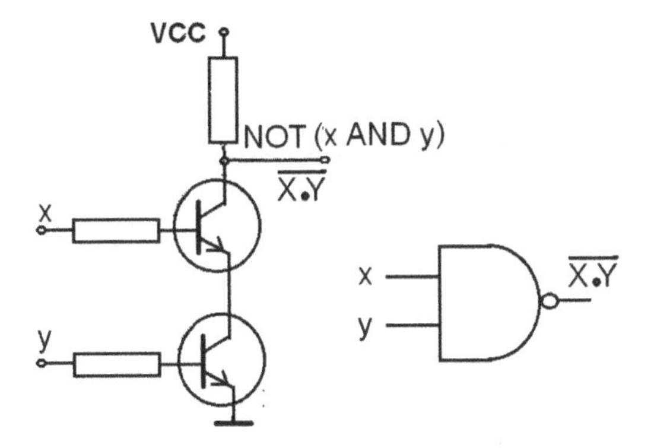


**Fig. 8.9** Logic NAND gate

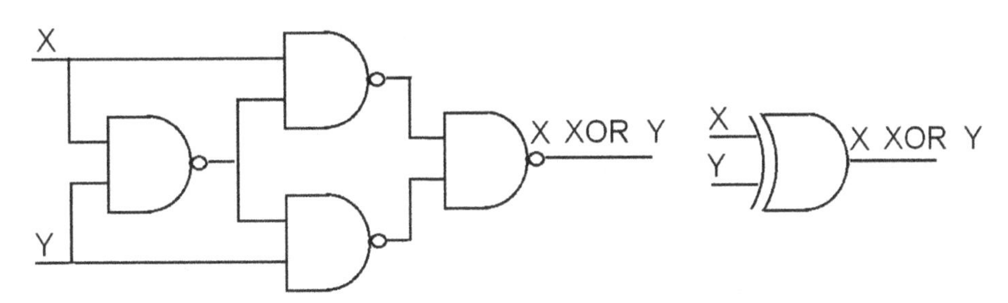


**Fig. 8.10** XOR gate

### 8.2.1 Enable/Disable Signal with AND Gate

Figure 8.11 shows an application of the AND gate to enable or disable signal train that comes from a clock generator or other source. The signal can pass through the gate only when the Y input is high. Note that this method can hardly be used to enable or disable analog, continuous signals like audio wave. The output of the gate can only be "0" or "1".

### 8.2.2 Clock Signal Generator

In Fig. 8.12 there is an example of a clock rectangular signal generator. It consists of two invertor gates and a RC circuit. (here "0" stands for low voltage, and "1" for high voltage).

Suppose that we start with voltage $V_1 = 0$, So, we have $V_2 = 1$, $V_3 = 0$. So, the capacitor C begins receive charge through the resistor R. It is a simple RC circuit, where the charging is an exponential curve, (the voltage $V_1(t)$ ). When $V_1$ reaches the high level (approaches state "1"), then the gate A switches to $V_2 = 0$, and $V_3$ changes to 1. With these voltages, the charging current of C inverts, and $V_1$ decreases. When it approaches zero, the output of gate A switches to 0, and all the process repeats. The changes of voltages $V_1$, $V_2$ and $V_3$ are shown in the figure.

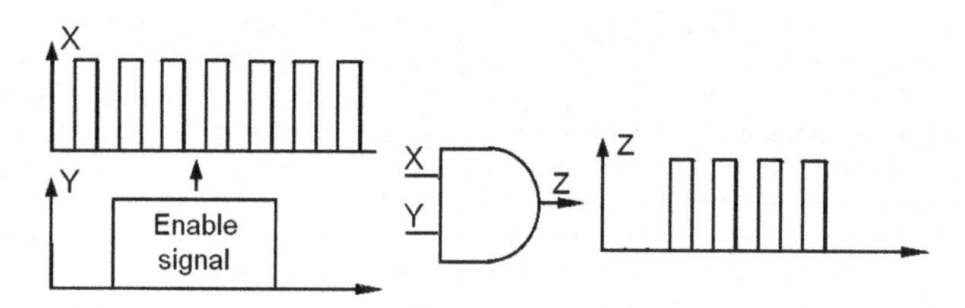


**Fig. 8.11** Using the AND gate to enable signal

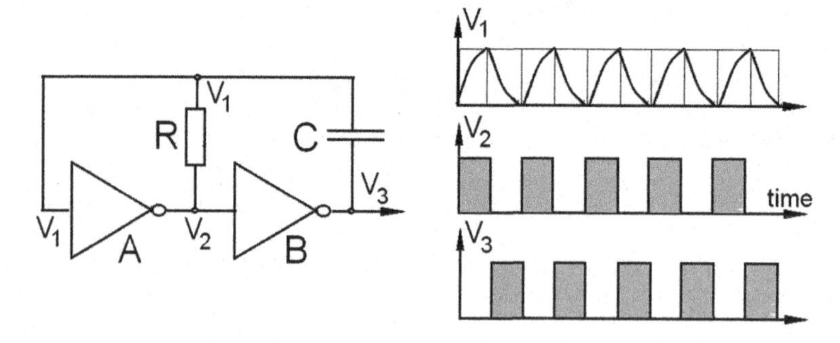


**Fig. 8.12** Clock signal generator

### 8.2.3 Flip-Flops

The flip-flop circuit constructed with transistors has been discussed in Sect. 4.3.3. Using logic gates, it is easy to construct flip-flops. In Fig. 8.13 we can see a flip-flop, constructed with NOR and NAND gates.

Consider the circuit of Fig. 8.13, part A. Suppose S = 0, R = 0, and the state of gate $g_1$, $Q = 1$, As the gates are of type NOR, the output of $g_2$ must be 0. This way, at both inputs of $g_1$ we have logical "0", and the supposed state "1" remains. Now, let us apply a "1" at input S. Now, $g_1$ has inputs "1" and "0", so it must switch to output "0" (it is NOR). Thus, $g_2$ receives inputs "0, 0", so it must have output "1". This state will remain also when the input S changes to "0". In a similar way, we can prove that when R becomes equal to "1", the outputs of $g_1$ and $g_2$ switch to "0" and "1", respectively.

The flip-flop of part B of the figure shows the diagram of the NAND-based flip-flop of type "$\overline{SR}$". The difference is that at the inputs we apply logic negations of signals S and R.

The flip-flop can also have the additional input "E" or "CK" that disables the changes of state. This allows to synchronize the device with other elements that use the same source of clock.

Among the variety of flip-flop types, an important one is the *D-type flip-flop* (Fig. 8.14).

The input D is the data input, where "0" or "1" are applied. CLK is called *clock pin.* If there comes a positive pulse to D, then the circuit checks the input D and may change state, provided CLK = 1. If CLK = 0, then nothing happens. When CLK changes to 1, then the next rising edge of CLK changes the circuit state to Q = 1. If the input D returns to 0 and

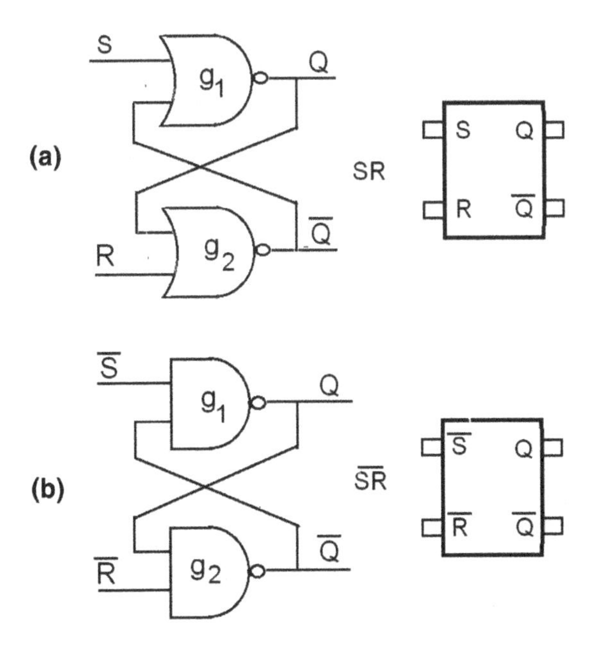


**Fig. 8.13** Flip-flops

**Fig. 8.14** D-type flip flop

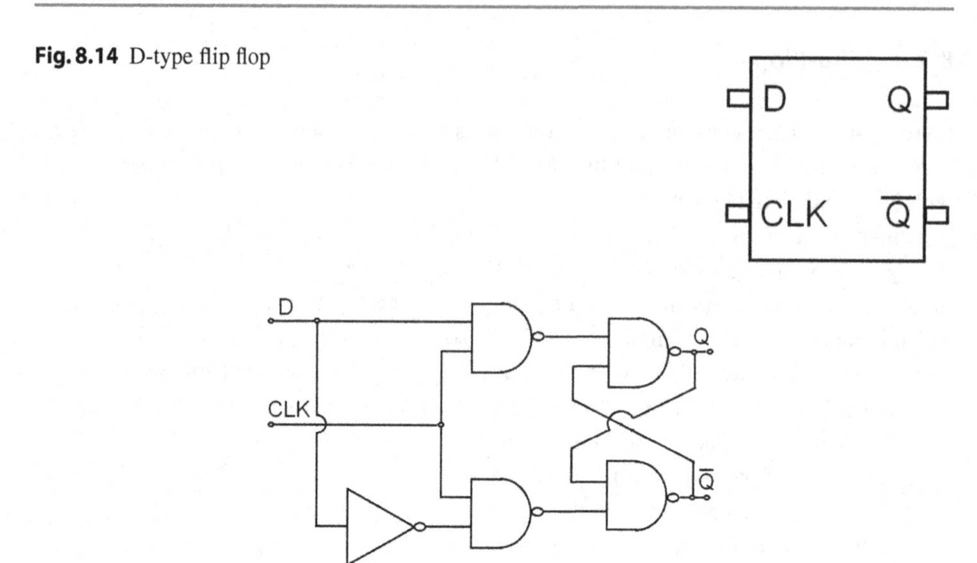


**Fig. 8.15** D-type flip-flop

the CLK does not change, then the state of the flip-flop does not change. At the output Q the last value of D persists, until there is the next rising pulse on CLK (Fig. 8.15).

### 8.2.4 Binary Counter

The D-type flip-flop can be used to construct a binary counter of received pulses. In Fig. 8.16 we can see and example of a 3-bit counter. It is composed by three flip-flops, where the output of each of them is the CLK input to the next one.

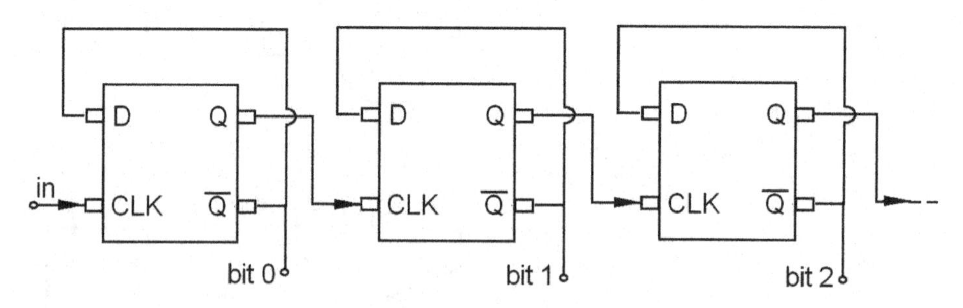


**Fig. 8.16** Binary counter

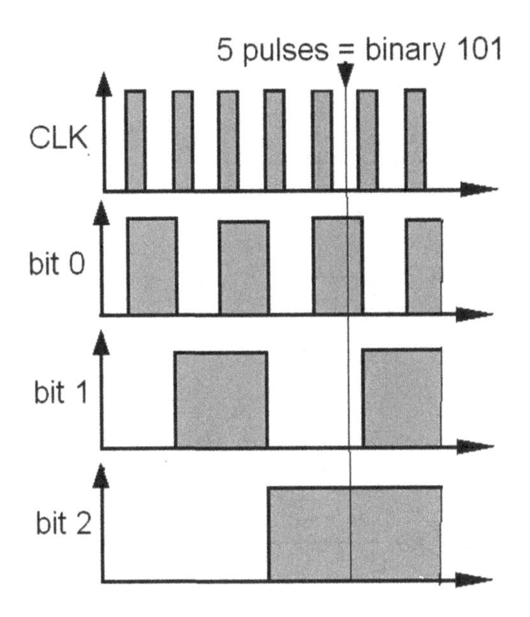


**Fig. 8.17** Binary counter, waveforms

Figure 8.17 shows the wave-forms of the counter, in response to a sequence of input pulses. Observe, for example that the counter state after five incoming pulse is equal to 101, which is the binary representation of five. Of course the input sequence to be counted need not be a regular periodic train of pulses. The pulses may be events distributed over time in any discrete time instants.

### 8.2.5 Binary Adder

One of the basic components of any digital data-processing device is the *binary adder*. Supposing that the machine uses binary number representation, the adder is composed by a series of adder stages, depending on the required resolution (32, 64 or other).

Each stage of the adder receives the two bits to be added, and the *carry out* that is passed to the next stage. As the result, the stage produces the sum bit, and the carry-out bit for the next stage. Figure 8.18 shows a possible diagram of one adder stage, using NAND, NOR, XOR and invertin (NOT) gates.

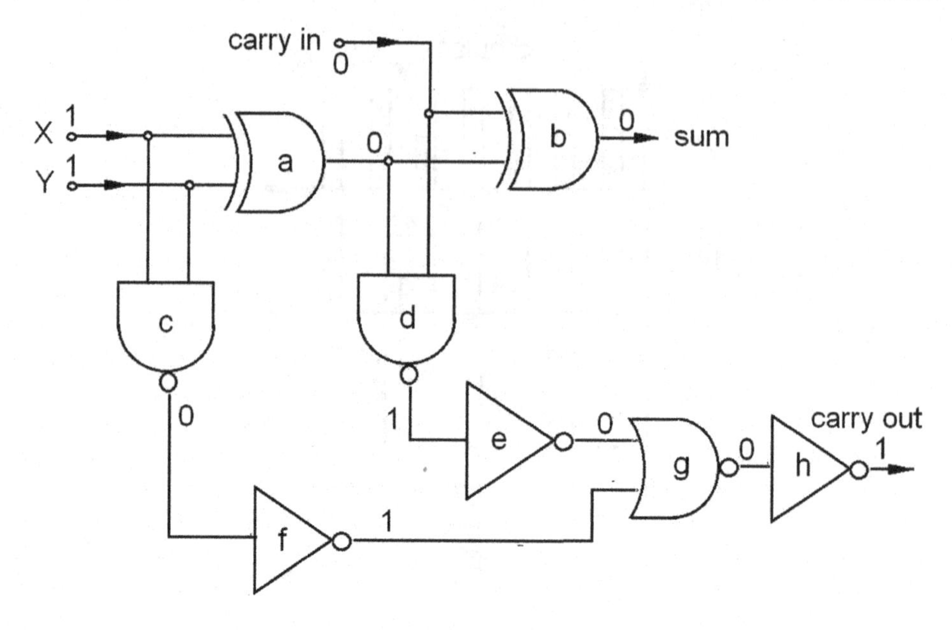


**Fig. 8.18** Binary adder, one binary position

In the Fig. 8.18 we can also see an example of adding binary numbers 1 and 1. The "zeros" and "ones" on the gate outputs illustrate the adding process. It can be seen that the sum is equal to zero, and the carry-out bit for the next stage is equal to 1. The whole true table for this circuit is as follows.

$$(X, Y) = (0, 0)\ carry\ 0 \rightarrow sum\ 0\ carry\ 0$$
$$(X, Y) = (0, 1)\ carry\ 0 \rightarrow sum\ 1\ carry\ 0$$
$$(X, Y) = (1, 0)\ carry\ 0 \rightarrow sum\ 1\ carry\ 0$$
$$(X, Y) = (1, 1)\ carry\ 0 \rightarrow sum\ 0\ carry\ 1$$
$$(X, Y) = (0, 0)\ carry\ 1 \rightarrow sum\ 1\ carry\ 0$$
$$(X, Y) = (0, 1)\ carry\ 1 \rightarrow sum\ 0\ carry\ 1$$
$$(X, Y) = (1, 0)\ carry\ 1 \rightarrow sum\ 0\ carry\ 1$$
$$(X, Y) = (1, 1)\ carry\ 1 \rightarrow sum\ 1\ carry\ 1$$

This way, we enter the field of digital computers. This is not the topic of this book, so this chapter ends at this section.

# Bibliography


1. Astrom, K. J., & Hagglund, T. (1995). *PID controllers*. Instrument Society of America. ISBN 1556175167.
2. Alexander, C. K., & Sadiku, M. N. O. (2009). *Fundaments of electric circuits*. McGraw-Hill. ISBN 978-0-07-352955-4.
3. Boashash, B. (Ed.). (2003). *Time-frequency signal analysis and processing: A comprehensive reference*. Oxford: Elsevier Science. ISBN 978-0-08-044335-5.
4. Ifeachor, E. C., & Jervis, B. W. (1993). *Digital signal processing: A practical approach*. Addison-Wesley.
5. Peyret, R. (2002). Chebyshev method. In *Spectral methods for incompressible viscous flow*. Applied mathematical sciences (Vol. 148). New York: Springer. https://doi.org/10.1007/978-1-4757-6557-1-4.
6. Thin, M. M., Than, M. M., & Myint, T. (2015). Enhancement of SAR algorithm for analog to digital converter. In *Proceedings of 2015 International Conference on Future Computational Technologies (ICFCT'2015)*, Singapore, March 29-30. ISBN 978-93-84468-20-0.


© The Editor(s) (if applicable) and The Author(s), under exclusive license to Springer Nature Switzerland AG 2023

S. Raczynski, *How Circuits Work*, Synthesis Lectures on Engineering, Science, and Technology, https://doi.org/10.1007/978-3-031-34934-8

# Index



© The Editor(s) (if applicable) and The Author(s), under exclusive license to Springer Nature Switzerland AG 2023

S. Raczynski, *How Circuits Work*, Synthesis Lectures on Engineering, Science, and Technology, https://doi.org/10.1007/978-3-031-34934-8

Printed in the United States
by Baker & Taylor Publisher Services